saving planet earth

What is destroying the Earth
and what you can do to help

Tony Juniper

An Imprint of HarperCollinsPublishers

978-0-00-726183-3

Based on the BBC television production Saving Planet Earth
© BBC 2007

978-0-06-154451-4 (in the United States)
FIRST U.S. EDITION Published 2007

HarperCollins books may be purchased for educational, business, or sales promotional use. For information in the United States, please write to: Special Markets Department, HarperCollins Publishers, 10 East 53rd Street, New York, NY 10022.

10 09 08 07
6 5 4 3 2 1

Color reproduction, printed and bound in Great Britain by Butler & Tanner

This book is proudly printed on paper which contains wood from well managed forests, certified in accordance with the rules of the Forest Stewardship Council. For more information about FSC, please visit www.fsc-uk.org

Mixed Sources
Product group from well-managed forests and other controlled sources
www.fsc.org Cert no. SW-COC-1806
© 1996 Forest Stewardship Council

FSC

contents

1.

living earth

unique earth

Our planet is utterly unique—a blue, white, and green gem suspended in the gaping black emptiness of space. The more we learn about the Earth and the rest of the cosmos, the more we realize that the chances of life occurring anywhere else are indeed slender. The dead planets that comprise the rest of our solar system bear silent witness to how vastly different our world is to the others that we have been able to explore either with telescopes or space craft. That life occurs here in such abundance and diversity is nothing short of miraculous.

Our near neighbors are very different. Mercury is seared by unbearably hot temperatures. The closeness of that world to the Sun renders life there impossible. The solar orbit of the planet Venus is more distant from the Sun (closer to the Earth's, as well as being closer in size to the Earth), but, with an atmosphere mainly composed of carbon dioxide and blazing surface temperatures of some 932 degrees fahrenheit, conditions there are impossibly hostile. Beyond the Earth is the orbit of Mars, from where there are some tantalizing signs that life might at least once have existed. However, even if that is so, it is now a cold, waterless world with a thin dead atmosphere that so far has not revealed conclusive proof that life has ever occurred there.

The gas giants of the solar system, Jupiter and Saturn, are chilled, hostile worlds. While some believe that life might dwell beneath the frozen surface and wastelands of Europa, one of Jupiter's moons, speculation has so far not led to evidence that this is indeed true. Uranus, the first planet discovered using a telescope, is so far from the Sun that its temperature remains below –328 degrees fahrenheit. Further out still, the orbit of Neptune is so distant that the Sun's light barely warms the planet, but its atmosphere of methane has the highest-known wind speeds in the solar system and near-supersonic gales tear at the planet's barren icy surface.

The Earth could so easily have remained like these other lifeless worlds. Because of its apparently perfect distance from the Sun, however, it transformed from a dead rock into the vibrant world that we know today, despite its isolated existence.

The realization that we live in a fragile and remote world was first brought home to us with the early space missions in the 1960s. Pictures brought back from the vast vacuum beyond our atmosphere showed the vulnerability and limitations of our small planet.

Bathed in blue oceans spawning white swirls of cloud that nurture the green expanses of land, it was obvious to all that our world is indeed special. However, in the half century since those first pictures of our world were brought back to Earth, there has been a progressively more certain view that what we are doing to that precious, unique, and finite world is causing changes beyond precedent.

When seen from the moon, the Earth is a small, fragile oasis of life in the desert of the universe.

the earth has supported life for over 3.5 billion years—the first recorded life was similar to stromatolites now found in Australia.

the birth of life

Life first stirred on this planet about three and a half billion (three thousand five hundred million) years ago. At first, and for immense periods from then on, it was comprised only of the simplest single-celled organisms. Powered by sunlight, these early life forms took carbon dioxide gas and water from their environment and used these materials to make carbohydrates as a means to produce energy. As a by-product they released oxygen. Over billions of years these single-celled organisms therefore changed the atmosphere, altering the relative proportions of oxygen and carbon dioxide—increasing the former and reducing the latter—and making it easier for other organisms to evolve.

Multi-celled organisms eventually appeared in the oceans, followed by creatures with hard shells. Rocks formed from deposits laid down about 600 million years ago contain fossils that show how an explosion of life occurred at about this time.

As oxygen levels in the atmosphere increased, so it was possible for life to expand on land too. There were not a number of continents like today; rather, there was a great super-continent comprising most of the world's land. The first plants were followed by the first creatures with primitive lungs. Later, luxuriant forests grew. Vast quantities of carbon were captured by the trees, and then stored away as dead trunks and branches were smothered by sediment and, over immense periods of time, encased in rock. During this period there was an explosion in the diversity of animals.

Then, life was nearly wiped out.

About one-quarter of a billion years ago, 95 percent of marine species and 70 percent of land families were suddenly eliminated. So severe was the impact of the changes that occurred at this time that the so-called end-Permian extinction nearly saw the end of life on Earth. There is no clear conclusion as to why this cataclysm overtook our planet, but a sudden climate change is among the more likely contenders.

Slowly, from the life forms that remained, natural evolutionary processes led again to a gradual diversification of plants and creatures. The age of reptiles arrived: giant dinosaurs roamed the land and swam in the seas; some even took to the air. It was much warmer than today—average global temperatures were as much as ten degrees higher than now—and forests even grew near the poles. Life continued to diversify, with birds and mammals appearing and the reptiles taking on ever more impressive form. The seas teemed with fish, turtles, and myriad colorful corals and crustaceans; the land was luxuriant with plant life and insects proliferated.

Then, again, there was an abrupt destruction of life: 65 million years ago a meteorite collided with the Earth. The six-mile diameter object crashed down in the present day Gulf of Mexico (if you look at a map you can still see the shape of part of the crater—the semicircular coastline on the west side of the Yucatán Peninsula). It exploded with the force of millions of nuclear warheads, spewing smoke and debris into the atmosphere. The Sun's light and warmth was blotted out; giant tsunamis rushed across the land; acid rain fell everywhere, killing plants and scouring the land. The dynasty of the dinosaurs, the animals that had dominated the earth for scores of millions of years, came to an end.

Some creatures survived, however, including mammals and birds that had lived in the shadow of the great reptiles from which they were both descended. Once more, and over millions of years, life recovered. Evolutionary changes led to an amazing diversification of species. The age of the birds and the mammals had arrived. Warm and cool periods came and went. Forests and deserts grew and shrank. Sea levels rose and fell.

Following several millions of years of these alternate warm and cool periods, during which much of the world was coated with a thick layer of ice, today we enjoy the milder conditions of an interglacial period. The most recent Ice Age ended about 10,000 years ago, a mere heartbeat in the life of the Earth. In the wake of that last cool phase, animals and plants abound, including many species that sat out the last Ice Age—on average some five degrees cooler than today—in patches of suitable habitat that remained.

The faint mark left by the Yucatán meteorite impact that caused the extinction of the dinosaurs. Although this impact caused the end of the age of reptiles, the speed of extinction was slower than that being experienced today.

the largest living organism on earth—
the giant sequoia.

the fabric of life

Looking around us now at the incredible richness of life on Earth, it is truly awe-inspiring to realize that what we see today is the result of the three and a half billion years of biological evolution. That long journey of life, of extinction and evolution, of rise and fall, has led to a proliferation of life on this small and unusual planet that is truly breathtaking. Indeed, it is likely that, until very recently, and with the rise of humankind, the diversity of life on planet Earth was at its greatest in its entire history.

This immense variety spans many different ecosystems: from coral reefs to polar tundra and from tropical rain forests to temperate grasslands. The multiplicity of natural systems in turn supports a colossal number of different species.

A recent estimate suggests that there are of the order of 13 million species living today (although some calculations have indicated that the total could be as many as 100 million). From small birds that spend nearly their entire lives in the air, to the giant sequoia trees of California, from tiny nematode worms in the soil to the hardy plants that live on high mountains, life is nearly everywhere.

The uncertainty of how many different life forms there actually are arises from the fact that we know surprisingly little about the fabric of life on our home planet. To date, and following more than two centuries of systematic investigating, cataloging, and describing, we have given names to between 1.7 and 1.8 million species. On the basis of what we know, the other millions we assume exist. From the canopies of the rain forests to the depths of the ocean, our planet still holds many surprises.

One study of insects in an area of rain forest canopy discovered that five out of six of the many different kinds were found to be new to science. But new finds are not just being made on land. It is estimated that nearly half of the freshwater fish of South America have yet to be described and named. And an exploration of deep-sea hydrothermal vents recently led to the discovery not just of new species, but of entirely new families of creatures.

Even among the primates, our closest cousins in the great branching tree of evolution, we are still finding hitherto unknown species. Even since the year 2000, several previously unknown ones have been unearthed. One was discovered in the amazing and unique mountain forests of Tanzania: with off-white bellies and tails, highland mangabeys are otherwise covered in dense brown fur, an adaptation to the primate's mountain habitat, where temperatures can drop below freezing at altitudes above 6,560 feet.

Another, a member of the macaque family, was found in the state of Arunachal Pradesh, which lies in India's remote northeastern region. Named the Arunachal macaque, this "new" monkey is a comparatively large brown primate with a short tail. The researchers who came across it were surprised to have found a hitherto-unknown large mammal in such a populous country. Perhaps this find is equivalent to stumbling across a new type of racoon in West Virginia, or an unknown type of fox in Scotland.

In Latin America, too, new primates have been found. For example, two species of Titi monkeys were recently discovered in the rain forests of Central and South America. Given that they are the size of domestic cats, it seems unlikely that they were missed— it is more likely that they were "new" because we haven't yet looked closely enough. Indeed, since 1980, about 40 "new" species of monkeys have been discovered worldwide, 13 of them in Brazil. It is a similar story for other groups too: birds, frogs, and plants included.

A team of scientists who recently visited an inaccessible part of the large tropical island of New Guinea in the easternmost and least-explored province of the western part of the island, found an astonishing mist-shrouded "lost world" that could only be reached by helicopter. In an area of highland forests of about 740,000 acres on the upper slopes of the Foja Mountains, rare and previously unknown animals and plants were found. They discovered dozens, if not hundreds, of new species in what expedition members said was probably the most unspoiled ecosystem in whole Asia-Pacific Region. The mountainous terrain there, and the long isolation from other major land masses, has permitted hundreds of distinct species to evolve, often specific to small areas. Even though they only had limited time, researchers still found: more than 20 new

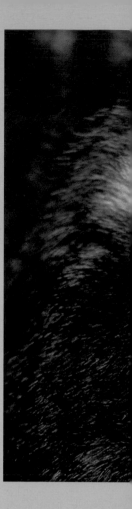

New species of large animals are still found to this day—40 new species of monkey have been discovered in the last 25 years, including a new cousin for these dusky titi monkeys. A new species of bat has also been found in the UK—the soprano pipistrelle.

species of frogs and 4 new butterflies; 5 new palms and what could prove to be the world's largest rhododendron; as well as many other plants completely unlike anything they had encountered before.

If we are still finding this many monkeys, birds, flowers, frogs, and butterflies, it seems that we must have even more to learn about the less obvious and not-so-charismatic creatures that share our world. While human attention is naturally attracted to the big, colorful, and eye-catching species, the myriad tiny plants and animals that enable our planet to function properly hardly seem to warrant a second glance. And, yet, it is these tiny living organisms that, more than any others, sustain life on Earth.

It may not be immediately obvious, but the functioning of our planet and its ability to support this staggering diversity of life is dependent on millions of different kinds of tiny single-celled plants and animals, a multitude of tiny worms, crustaceans, and

poaching and habitat destruction are driving the Sumatran tiger to **extinction**.

insects, and a mind-boggling array of different fungi. Between them, this vast workforce of tiny life forms undertakes the gargantuan task of recycling the Earth's organic waste; they release nutrients, enrich soils, and enable the plant growth upon which all animals ultimately depend. They help determine the composition of the Earth's atmosphere and are themselves at the base of many of the food webs that enable higher life forms to exist at all.

The diversity of ecosystems and species are not the only measures of our planet's incredible variety, however. Included within that 13 million or so different species that we believe share our planet with us are a vast number more of genetically distinct varieties. For example, there are six subspecies of tiger. Because these huge forest feline carnivores have become isolated from one another in different geographical areas, they have evolved distinct characteristics: the immense and powerful Bengal Tiger is found from India to Burma and north to Bhutan; the smaller and darker Indochinese Tiger occurs from Burma to Vietnam; the Malayan Tiger lives on the Malay Peninsula; the Sumatran Tiger, the smallest of all, is confined to the Indonesian island of Sumatra; the large pale and thick-coated Amur Tiger is found in Siberia; and the rarest, the South China Tiger, now no longer found in the wild, was once widespread in China. All of this variety is found within one species. Geographically isolated varieties of many other animals and plants show similar or even greater diversity. On top of all of this are the genetic differences that exist between individuals.

The tapestry of life on Earth is indeed rich and varied: collectively, this great global living treasure is called biodiversity. The sphere of life and the web of connections that bind it together is called the biosphere.

the strawberry poison dart frog is one of the 1770 of the 5743 known species of amphibian that are currently at risk of becoming **extinct.**

life under threat

This miraculous and vibrant manifestation of life is, however, disappearing before our very eyes. Since we haven't yet built up a full inventory of species, let alone subspecies, it is impossible for us to accurately state the rate of loss now taking place. But biodiversity is certainly being depleted on a massive scale.

Although the Earth's history shows that there can be many natural reasons for the loss of biodiversity, the principal agent at the root of this particular decline is undoubtedly the collective actions of humankind.

For millennia our species has contributed to the extinction of other species. For example, much of the so-called Pleistocene megafauna, including huge mammals such as mammoths, wooly rhinos, giant ground sloths, and huge armadillos, were most likely wiped out by prehistoric humans in combination with, and possibly hastened by, climatic changes in North and South America, Europe, and Australia. The extinction of these creatures was a tragic loss. But the speed at which species (never mind distinct populations) are now disappearing has gained enormous momentum and is certainly having a bigger impact on life on Earth than anything witnessed in the planet's recent past.

One estimate suggests that, within about the next hundred years, one half of all species could be extinct. Indeed, a poll of leading biologists revealed that the majority of experts believed that the world is in the midst of a mass extinction comparable to the dramatic declines of life etched in the fossil record, such as the end-Permian catastrophe, or comparable to the event which marked the end of the dinosaurs. Five such major extinction episodes are indicated in the fossil record, but the sixth—the one that now seems to have been initiated by humankind—is going faster than any one that has gone before.

The full effect of previous large-scale extinctions, even those caused by an instantaneous event, such as a meteorite collision, was not felt for the equivalent of many human lifetimes, due to the complex ecological interactions that were unleashed. Other periods of rapid species loss took many thousands of years, or even longer, to gather the pace of the loss taking place now.

It is true that species are always being lost; as other species evolve and conditions change, some quietly slip away. This is

irrespective of whether some major planet-wide change is taking place and, certainly, this is a part of the normal functioning of natural systems. But this is only one small aspect of what now appears to be happening. Indeed, at the present assumed rate of species loss, extinctions are taking place at between one hundred and one thousand times faster than that average or "background" level. Moreover, the current rate of extinction has even been calculated to be proceeding at between ten and a hundred times faster than any of the prior mass-extinction events in the history of the Earth. If these estimates are correct, then the pace and scale of change now taking place is without precedent.

Once the loss of biodiversity gathers speed, it creates a certain momentum, and accelerates as relationships between species begin to break down. For example, the loss of a pollinator, such as an insect or bat, can doom the plant it pollinates, and a prey

The wooly rhinoceros is one of the many species of megafauna that is thought to have been driven to extinction by humans— through habitat destruction and hunting.

species can take its predator with it into extinction. One estimate suggests that every tropical plant species that becomes extinct takes some 20 insects to oblivion with it.

The number of documented extinctions so far does not seem very high, however, at least when compared to the total number of species that are estimated to be here on Earth with us. According to a thorough 2004 estimate, at least 784 species have become extinct since the year 1500. The majority of these lived on land, others in freshwater with the remainder from the marine environment.

The most up-to-date list of those threatened with extinction is 15,589, which includes one in four mammals and one in eight birds. The total number of known threatened animal species increased from 5,205 in 1996 to 7,266 in 2004. The group that is most threatened of all is the amphibians: some 1,770 of the 5,743 known species are believed to be at risk of being lost—that's 31 percent.

While, at first glance, these figures for the total number of actual documented extinctions and the number of immediately threatened species may not seem that high, the expected rate of extinction over the course of this century should send alarm bells ringing planet-wide. Experts already put the real rate of loss at about one species per hour, and expect this to accelerate.

Less obvious than total figures, and less well documented, is the disappearance of unique populations and subspecies. Each genetically distinct population of animals or plants has its own characteristics: some of these might be visible, such as differences in size or color; others might not be so obvious, such as the ability to withstand cold or drought.

Of an estimated one to six billion different populations on Earth, a conservative calculation based, for example, on known rates of habitat loss concluded that populations are disappearing in tropical forests at a rate of 1,800 populations per hour.

Once the Earth's biodiversity is gone, it cannot be brought back. It will be gone forever. The long historic threads of our planet's biological evolution over billions of years will thus be slashed in, what is for the life of the Earth, a mere blink of an eye.

habitats in decline

While biodiversity is often measured and expressed in numbers of species, it is habitats that are the critical factor in determining the health of life on Earth. From the depths of the oceans to the exposed peaks of high mountains, it is the well-being of natural environments that is most critical to the conservation of species and populations of animals and plants.

on land

The most biologically diverse ecosystems on Earth are the tropical rain forests. In a broad green swathe around the equatorial regions that covers just six percent of the world's land area, these lush, moist formations contain perhaps half of all of the world's land species. Even within this relatively small proportion of the planet are so-called hotspots, where species richness is especially high. For example, the area containing the highest plant diversity on Earth, with over 40,000 species, is concentrated in just two percent of the Earth's surface in Colombia, Ecuador, and Peru.
Of the quarter of a million or so vascular plants to have been defined, some 70 percent grow in the tropical rain forests. Compared to seasonal temperate forests or the colder boreal forests, an incredible level of diversity is found. In a 24-acre plot of Malaysian rain forest, there are 780 tree species—more than the total number of tree species native to the whole of the USA and Canada. In a survey of forest near to the Peruvian city of Iquitos, 300 different kinds of tree were found in just one 740-acre patch of forest. And just 250 acres of Amazon rain forest can contain up to 1,500 different plant species—as many as in the whole of the UK.

Even more notable than the mind-boggling variety of plants found in tropical rain forests is the staggering diversity of insects. A Peruvian scientist discovered over 1,200 species of butterfly in the 13,600-acre Tambopata reserve. Beetles are found in even greater profusion: in some tropical forests it is estimated that some 20,000 species inhabit 2.5 acres, while, in the whole of the USA and Canada, there are just 24,000 known species.

Incredibly, this amazing richness of plants and animals is sustained on poor, often sandy, soils. The powerhouse that sustains such systems is the Sun. Like a great shimmering machine, the rain forests use energy from light to manufacture stems, leaves, flowers, and fruits through the incredible process of photosynthesis. As a by-product, vast quantities of water vapor are exhaled. As it rises to cooler altitudes, the clouds condense and give birth to great, crashing thunderstorms. The soils remain impoverished, however, because the forest nutrients are in the plants, not the ground. The constant year-round tropical growth also means that demand for nutrients runs all year and plants thus immediately take back organic matter that falls to the forest floor as it is recycled by a host of bacteria, fungi, termites, and other organisms.

Some of the nutrients that enable this mighty equatorial biological powerhouse to function even come from North Africa. Observations from space have revealed that the strong harmattan winds, which in winter whip across the southern Sahara, focus into one tiny area called the Bodele depression. Between two mountain ranges, the wind is forced upward into the atmosphere. Fine dust particles are blasted skyward, providing the largest single source of dust on Earth: some winter days three-quarters of a million tons rises into the air, with about fifty million tons per year wafted across the Atlantic to fall on the Amazon forests some three thousand miles distant. Amazingly, one of the world's wettest ecosystems is thus directly dependent on one of the driest to function properly.

The vast basin of the Amazon has the largest block of rain forest on Earth, and it is drained by world's largest river. Its annual outflow accounts for one-fifth of all the fresh water that flows into the world's oceans; where the Amazon River reaches the Atlantic it is as wide across its mouth as the distance from London to Paris.

This vast ecosystem of forest and rivers is the largest repository of biodiversity on Earth, containing a high proportion of the planet's total biodiversity. The Amazon basin forests are crucial for the wider planetary system in other ways too. For example, locked away in plants, the relative balance of carbon in trees compared to the air is, as we shall see in the next chapter, an important part of planetary stability.

But all is not well in the Amazon, or indeed many other of the areas of the tropical rain forests that remain.

One satellite photograph taken in September 1987 showed a total of 7,603 fires burning in the Amazon river basin rain forests. This image and many others like it sparked international concern and stimulated a plethora of initiatives to stem the destruction. Although the rates of forest loss have alternately declined and increased, the continuing loss of these rain forests remains a cause for serious global concern today.

The reasons for forest loss across this vast region are many and varied. Logging, both legal and illegal, takes a serious toll. Logging operations also open up the forest with roads that allow further clearance for farming (both small- and large-scale). Mining operations also contribute, and have an impact both on tree cover and, in many cases, water quality; serious pollution has been linked to many mines. The clearance of large areas of forest for beef production and for soya have also contributed significantly to deforestation. Global demand for meat on the one hand, and for animal feed on the other, have helped boost these industries across Amazonia in recent years. Indeed, around the world, it is the expansion of agriculture that is having the most serious effect on the remaining tropical forests.

In Indonesia the rain forests are falling to the march of palm-oil plantation expansion. Today, palm oil is grown on an ever more huge scale, providing global commodity markets with vast quantities of cheap vegetable fat. Across the Indonesian islands of Sumatra and Borneo, palm-oil plantations have so damaged the rain forest that experts expect the extinction of the orangutan in the wild by about 2020, if nothing is done. More than 90 percent of the orangutan's original habitat is gone and the remainder is under grave pressure, with the palm-oil industry being backed by the Indonesian government even in protected areas where the last orangutans live, for example, in southwestern Borneo. The forests on these islands are also the home of countless other unique, rare, and declining species.

The orangutan of southeastern Asia—close relative of humans—is also threatened by habitat destruction and hunting.

Logging operations and conversion to plantations can increase the risk of serious fires, especially when coupled with unusually dry conditions. Large-scale conflagrations in turn lead to further forest loss and may increase pressure on adjacent virgin forests by improving access to formerly remote areas. They also cause major public health problems across Indonesia and Malaysia, as the haze of smoke drifts across urbanized areas.

Alongside government-backed settlement programs or conversion of forest for large-scale ranching or agriculture (common causes of deforestation in Latin America and Asia), another explanation for forest loss is the spread of subsistence farming (more common in Africa and Asia). The impact of small-scale forest conversion for subsistence needs can be quite different to large-scale clearance in terms, for example, of soil erosion and the wider functioning of the forest regional or microclimate.

The rapid loss of rain forests across the southeastern Asian region, for whatever reason, is driving many species to the edge of extinction—and some have already gone over it. But what is now taking place across the tropics is not, in a global sense, a new trend. In many European countries, and in parts of Asia, such as China, the process of deforestation has been in progress for some time, in many cases for several millennia. In the New World tropics, lowland, seasonal, deciduous forest began to disappear after 1500 with Spanish and Portuguese colonization, because these were the regions most easily converted to agriculture, not least because of their more manageable climate. The rain forests themselves came under attack mainly in the twentieth century and it is this assault that is leading to larger-scale impacts on biodiversity than that which accompanied the loss of some other forest types. This is because of the very high species diversity and the fact that many life forms found in these systems are either highly localized or highly specialized in their ecological needs—or both.

Compared to the year 1800, only about half of the tropical forests that then existed now remain. Asia lost almost one-third of its tropical forest cover between 1960 and 1980—the world's highest rate of forest clearance; almost 90 percent of West Africa's

Upper Guinea rain forests have been destroyed. In all, between 1960 and 1990, about 1.7 million square miles (over 400 million hectares) of tropical forest were cleared, an area roughly equivalent to half the size of the USA.

The current rate of rain forest loss is estimated at near two per cent annually (that's about 39,000 square miles (25 million acres) destroyed, with another 39,000 square miles being degraded). While there is inevitably huge uncertainty regarding the present rate of loss, and no knowledge of what it will be in the future, it is quite plausible that tropical forests will be reduced to between 10 and 25 percent of their original extent by around 2100. The impact of that scale of loss of the Earth's biodiversity would be extremely serious.

The state of the world's forests is not, however, simply a matter of how much there is. It is increasingly clear that the quality of forest is also vitally important. Some evaluations of forest status include plantations alongside natural forest cover in assessing rates of total forest loss or increase but from a conservation perspective, dense plantations of, for example, exotic conifers, can certainly not be regarded as compensation for the loss of natural rain forest.

Even where natural forests are left alone following some damaging activity, important issues of forest quality need to be understood in making judgements about forest value for biodiversity conservation. Quality measures might relate to tree health, forest biodiversity, and the age profile of the habitat (such as the number of mature compared to young trees). For example, selective logging of natural forest is not generally counted as deforestation, since logged-over areas can, in theory, regrow to fully functioning forests. Even if this might ultimately occur, in the short to medium term, logging degrades forest quality, leading to soil and nutrient loss and damaging the habitat of many species.

A large proportion of the tropical rain forests that do remain are not top quality from the perspective of the once natural biodiversity, and this is set to worsen. Logging pressures in many of the remaining large, virgin rain forest areas continue to intensify, with logging activities shifting from the heavily deforested areas of southeastern Asia, West Africa, and Central

America to the still largely pristine rain forests of the Amazon basin, Papua New Guinea, and central Africa.

The loss and degradation of tropical rain forest is thus a vitally significant cause of known extinctions. As the process continues during the twenty-first century, it is set to become perhaps the main cause of mass extinctions arising from human activity.

It is of course not only the tropical rain forests that are hemorrhaging biodiversity, however. Other forests—from subtropical to temperate to high-latitude boreal forests—are, to different extents, also shedding species and populations as they are cleared and degraded.

Grasslands and heathlands are also falling before the onslaught of farming, urbanization, and pollution. Even agricultural landscapes that, until recently, maintained some level of biodiversity in field margins, watercourses, and small fragments of seminatural habitats are suffering further depletion of biodiversity as farming methods become more and more intensive.

The present global scale of change to natural systems is utterly without precedent in human experience. A recent "stock take" on the state of nature reached some remarkable conclusions. The Millennium Ecosystem Assessment, prepared by 1,300 experts from 95 countries, concluded that humans had changed the world more rapidly over the last 50 years than during any other time in history.

The Assessment found that more land has been converted to cropland since 1945 than during the whole of the eighteenth and nineteenth centuries combined; approximately one-quarter of the world's coral reefs and about 35 percent of the mangroves in the countries surveyed were destroyed or badly degraded in the last decades of the twentieth century. This rate of change to natural systems is clearly unsustainable and is contributing to large-scale biodiversity loss in most of the Earth's main habitat types. The rate at which agricultural land is expanding varies from region to region, but much of the agricultural conversion taking place now is located in Latin America, sub-Saharan Africa, and south and southeastern Asia—precisely those areas in which most biodiversity is located.

It is not only the extent of the loss or even the quality of what remains that is the key determinant of ultimate extinction levels for many ecosystem types. Fragmentation is also a very major issue. Small and isolated areas of wetland, forest, dunes, or grasslands are rather like islands. If a particular species is lost due to disease or drought, for example, then it is far less likely that it will be able to recolonize from another area of suitable habitat if the area surrounding the patch from where it has disappeared is intensive farmland, roads, or urbanized. This is especially true if the species in question has low powers of dispersal; animals that can't fly and plants that produce seeds that don't travel very far are, of course, at particular risk. The loss of a species from individual patches of habitat over time could lead to complete extinction (as well as hastening the loss of genetically distinct populations in the process).

Species that require large home ranges will also be especially vulnerable as suitable habitat becomes progressively more fragmented, and may not survive in parts of their range because they simply don't have enough room. Large carnivores are especially at risk—grizzly bears and tigers, for example. Maintaining viable populations of these animals, in the hundreds at least, in order to permit a full genetic transfer between groups, requires large areas of territory.

Smaller areas of natural habitat are also strongly affected by their surroundings, especially in relation to microclimate or even regional climate. There is already evidence that small chunks of rain forest are affected in this way: in Amazonia, it is expected that the present process of rainfall generation created by solid forest will begin to break down with forest fragmentation, and the forest as a whole will thus become progressively drier. This, in turn, will ultimately affect the whole system as it will increase fire risk, potentially leading to the wholesale degradation of the forest and its biodiversity.

Development pressures can also lead to important losses of habitat, especially where they are already fragmented or where small areas of habitat play an important role in conserving particular species. For example, several kinds of endangered sea

turtle have been placed at greater risk of extinction because their breeding routines are disrupted by tourist hotels built near to the beaches where they lay their eggs. The Grenada dove, a member of the pigeon family that is unique to that island, is similarly at risk. Fewer than 180 of these birds exist, yet one of its most important strongholds, the Mount Hartman National Park, is under threat from a development of a tourist hotel and golf course. This could be the final blow for a unique species of bird that is already reduced to perilously low numbers.

Looking into the near future, it is clear that combined effects of

Papua New Guinean tribal headdress. Both the indigenous people and wildlife are threatened by the continual destruction of their habitat.

habitat loss, degradation, and fragmentation constitute the most important factors in hastening the extinction crisis that is unfolding around us on land.

the marine environment

It is not only the decline of forests and other terrestrial habitats that has serious consequences for biodiversity. In the marine environment too there is large-scale change taking place, and particularly symbolic of the threat to that environment is perhaps what is happening to the world's coral reefs.

These amazing marine systems are often compared with the tropical rain forest for their incredible diversity, and with good reason, since they are among the most diverse ecosystems on Earth. And, in common with the tropical rain forests, scientists still routinely discover new species: of fish, corals, worms, mollusks, crustaceans, urchins, and others.

In some ways, the corals that make up the structure of these systems are equivalent to the trees that provide the main fabric of a rain forest. Approximately one-quarter of all marine species inhabit coral reefs; the number of individual species may be as high as one million. The greatest variety of coral is found on reefs in the area in which the tropical Pacific and Indian Oceans meet. In the Philippines there are some 400 known species of hard coral, though, with increasing distance from this region, coral diversity declines. There are, however, still an incredible 350 species of hard coral in the northern part of the Great Barrier Reef off the northeastern coast of Australia, and about 250 further east into the Pacific Ocean around Fiji.

Corals are formed from the close association between animal polyps and tiny algal plants that live with them. To thrive, both need clean water at the correct temperature, and the algae need plenty of light in order to photosynthesize. When this relationship breaks down coral bleaching can occur. This is a process in which the algae are expelled by the polyps, the result being that the health of a reef can be dramatically harmed. Corals can recover, if

normal conditions return, but, if the stress continues, then reefs can die off completely.

Both of these basic conditions can be drastically affected by activities carried out many miles from the coral reefs. Indeed, the major threat to many of the world's coral reefs is the discharge of pollution and sediments from the land. This can include sewage water, agricultural pesticides and fertilizers, and soils that are swept into the sea as forests are cleared away or as farmland is washed away by heavy tropical rainfall. Farm fertilizers and sewage can set off the rapid growth of algae suspended in the

Coral reefs and rain forests are the two habitats where more species live than any other habitat on Earth. They are also the most delicate. The destruction of these two habitats results in the possible extinction of more species than any other habitats in the world.

water, thus cutting down the light reaching the seabed. Nutrients can also increase the growth of animals such as sponges, in turn depleting the space available for the corals to grow. Too much nutrient can also favor the increase in the number of Crown-of-Thorns starfish. These animals eat coral and, if their numbers reach very high density, they can inflict severe damage on reefs.

Power plants on land can cause large-scale damage to coral reefs. The periodic discharge of extremely hot water changes the water temperature, leading to coral bleaching and reef death. Development and over-fishing take a heavy toll as well. The removal of large numbers of reef fish has not only directly reduced the diversity of coral-reef communities, it has also, in some places, caused the coral-reef ecosystems to become unbalanced and allowed more competitive organisms, such as algae, which were once controlled by large fish populations, to become more dominant. This can upset the whole system, leading to a much

The crown of thorns starfish is an aggressive predator of coral.

larger loss of species than simply the commercially valuable fish.

As overfishing has led to reduced catches, fishermen have been forced to change their methods in order to sustain their income. The change of methods has included fishing nets and traps with smaller mesh that catch even the juvenile fish, thus causing even bigger population crashes as fewer and fewer of the fish make it to adulthood and breeding age. In some places, fishermen have resorted to the use of explosives and poison. These not only kill all fish (and a lot else besides), they severely affect the long-term health of the reef as a whole. Boat and ship anchors also cause widespread damage to reefs.

Some reefs are also plundered for corals for use in making jewelry or other decorative products. This can cause widespread reef damage, not least because the healthiest specimens are the most prized.

A wide range of other marine habitats are, of course, at risk. Sea defenses, the development of large ports, and the expansion of large coastal cities all contribute to the intense pressure put on coastal marine habitats. Structures such as groins and jetties have interrupted important long-shore sediment movements, thus causing marshes to become eroded.

And, as is the case on land, development can also affect particular threatened marine species. For example, oil and gas development on the island of Sakhalin off the far-eastern coast of Russia threaten the extinction of the Western Pacific grey whale. Already reduced to tiny numbers, the noise and disturbance caused by drilling and production could be the final nail in the coffin of another unique life form.

Intense pressure is now placed on many natural habitats and ecosystems. This is an almost planet-wide phenomenon, on land and sea. The good news is that, in most cases, the pressures can be reduced or stopped, and a high proportion of the biodiversity that remains can be saved. The bad news is that loss, degradation, and fragmentation of habitats is not the only factor driving the disappearance of the world's unique wildlife.

unwelcome visitors

An important, and often overlooked, cause of the depletion of biodiversity is the impact on native species from introduced ones. Animals and plants, which, in an ecological sense, are very aggressive, can wipe out more specialized creatures or plants with which they come into contact. Generalist creatures, like rats, can have an especially dramatic impact; prolific breeders and omnivores that eat more or less anything that is edible can devastate wildlife that has not adapted to either compete with them or to avoid being eaten by them; cats can have a similar impact on small birds and mammals.

Of all the documented extinctions to have taken place since 1600, introduced species appear to have played a role in at least half. The disappearance of many island species bears testament to how quickly entire life forms can be wiped out through inadvertent (or even deliberate) release of non-native species: over 90 percent of documented extinctions of amphibians and reptiles, and a similar proportion of birds, were of species confined to islands, and many were following the introduction of non-native species.

One of the most extreme examples is the fate of the Stephens Island wren. Stephens Island, a few miles off the coast of New Zealand, was the last refuge for this species. Although it seems once to have occurred on the main islands of New Zealand, it was wiped out there following the arrival of the Maori people, and more particularly the Polynesian rats that came with them. Toward the end, it only remained on Stephens Island. This last population has been said to have been wiped out by a single cat called Tibbles, who was brought to the island by its lighthouse keeper. Though this may be a myth, what is clear is that this unique bird was destroyed by feral cats that probably descended from an introduced, escaped domestic cat.

The wren was one of only a handful of flightless songbirds ever to be described (all of which lived on islands, and all of which are now extinct). It was also nocturnal. But so were the cats. The last refuge of the Wren proved no refuge at all, when shared with a cat. From 1894 onward, no one would have the opportunity to study these birds any longer. They were all gone. Amazingly, 1894 was

the dodo of the Northern Hemisphere—
the great auk.

not only the last year that these unusual and unique birds were seen, it was also the year that they were first described. They were wiped out in a matter of months.

The animal many see as the very symbol of extinction also ultimately fell victim to creatures introduced to its island home. The Dodo, native to the Indian Ocean island of Mauritius, was under pressure from deforestation and from being collected for food by sailors from passing ships. What probably completed the process, though, was a combination of pigs, monkeys, rats, and cats set free on its native island by visiting European vessels. This remarkable creature finally fell into the abyss of extinction in about 1680. Similar birds that lived on the neighboring islands of Reunión and Rodriguez disappeared shortly afterward.

Introduced species have had a dramatic impact on aquatic habitats too. The African Great Lakes—Victoria, Malawi, and Tanganyika—hold an incredible diversity of unique cichlid fishes. In Lake Victoria, however, the huge and voracious Nile Perch has become established and this single, exotic species may ultimately cause the extinction of most of the native fish, by feeding on them. This case is perhaps made all the more tragic because the Nile Perch was an intentional introduction to provide food and sport for people.

In many parts of the world, lakes and waterways have seen their native wildlife severely encroached upon by introduced zebra mussels. These fast-breeding bivalves can alter the abundance of algae and can affect nutrient levels in water, which, in turn, has a profound effect on whole ecosystems. It seems that no habitat or ecosystem is immune from attack from outsiders.

The loss of many other animals has also resulted from deliberate introductions, for example, that of a large proportion of the snails that were once unique to the Hawaiian islands. In 1955, in an effort to control the previously introduced giant African snail, a species of predatory snail from Florida was introduced. These voracious animals follow the mucous trails left by other snails, climbing trees to pursue their prey. Some of Hawaii's most vulnerable native snail species take an unusually long period to reach sexual maturity, and also have a low reproductive rate, thus

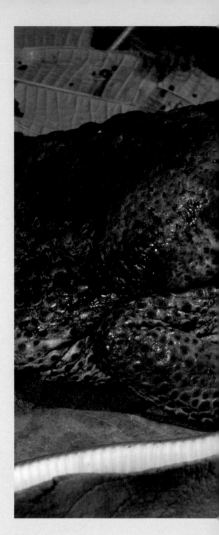

Cane toads have insatiable appetites and can reduce the population of local fauna very quickly

rendering them extremely vulnerable to extinction. Since the introduction of this foreign predator, some 15–20 native species of snail have gone extinct.

australia and new zealand

It's not only on islands, however, that introduced species can cause serious ecological disruption. Even on a continental scale, massive environmental damage can result. Just under 200 million years ago, Australia split away from the ancient super-continent that was once the Earth's main landmass during the Jurassic era. This ecological detachment resulted in many unique life forms being found there and nowhere else today. For example, while the ancient marsupials disappeared across most of the rest of the world because of competition with more modern animals, they survived on the isolated island continent—at least they survived until the arrival of Europeans. With human visitors came a whole host of other creatures that are now at large in the wild in Australia: these unwelcome visitors include camels, cane toads, goats, foxes, deer, rabbits, pigs, cats, dogs, and horses. Collectively, these newcomers have wrought massive ecological damage: they compete with native animals for food; they prey upon native animals; they damage native plants; and they degrade natural habitats.

Australia's native rabbit-eared bandicoot needs a constant supply of carbohydrate-rich seeds and roots to survive, but these animals must now compete with non-native creatures to meet this basic need. Rabbits degrade vegetation that provides food and shelter; foxes that were introduced to control the spiraling rabbit populations have killed native wildlife in at least equal measure to their intended quarry. Feral pigs destroy the vegetation that prevents erosion and provides food and nesting sites for native wildlife. They also compete with native animals for food and pose a serious threat to ground-nesting birds; by making muddy wallows, pigs have destroyed breeding sites and degraded key habitats of the endangered white-bellied frog, orange-bellied frog, and corroboree frog.

Feral cats, found in most habitats across Australia, have caused the extinction of some species on islands and have contributed to the disappearance of many ground-dwelling birds and mammals on the Australian mainland. They not only kill and eat native animals but also carry diseases that can be transmitted to them. On Macquarie Island in the southern Pacific Ocean, for example, feral cats caused the extinction of a subspecies of the red-fronted parakeet. On the mainland, they have most likely hastened (if not caused) the extinction of many small- to medium-sized mammals and ground-nesting birds and seriously affected the populations of bandicoots, mala, and numbats; they have also threatened the success of recovery programs for endangered species.

Another ecologically disastrous introduction into Australia was that of the cane toad. First released into Queensland in 1935, cane toads were brought to Australia from South America so as to help control cane beetles, which had proved to be a serious pest against sugar-cane cultivation. But the toads neither controlled the population of the beetles, nor did they stay in the sugar cane fields. They thrive in a wide variety of habitats, including urban areas, and they spread like wildfire, blazing a trail of destruction. The toads have no natural enemies, but the toxins they contain seriously affect the native animals that normally include frogs in their diet. With a voracious appetite and ability to breed quickly, cane toads soon began to out-compete native species of amphibian to the point where some were at risk of extinction. Despite several attempts to control cane toads, their inexorable march into new areas continues.

It's not just non-native animals that can cause problems. Plants and introduced-plant diseases can have a disastrous impact on native flora. Across many regions of Australia the native vegetation (and thus the habitat of native animals) is threatened by an aggressive species of root fungus. It was probably introduced into Australia with European settlement and has spread to take effect on hundreds of thousands of acres of native vegetation via trade and human migration. It is now a major threat to rare and endangered species.

New Zealand, also long separated from other land masses, similarly has animals and plants that are utterly unique. The islands' flora and fauna has thus been similarly vulnerable to the impacts of introduced species. New Zealand broke away from the ancient super-continent before land mammals had appeared and had been isolated for such a long time that it had no native land mammals (a couple of species of bat making up its entire complement of naturally occurring mammal species).

Birds reached the islands, however, and over long periods evolved into many unique forms. Among those that evolved were the many species of flightless moas. Without any mammals to compete with, or from which to fear predation, these birds took on the ecological niches that in other parts of the world would typically be occupied by furry rather than feathered creatures. With the arrival of human settlers from Polynesia, these birds were soon all extinct; the animals that the Maori people brought with them, including rats, also began to have an effect on the native creatures. Things worsened when Europeans arrived with their own many furry fellow travelers.

Many of New Zealand's unique birds, having evolved in the absence of land mammals, were flightless. This of course rendered them especially vulnerable to the army of introduced creatures that rapidly spread across their island homes. The animals that came with humans included rats, dogs, opossums, pigs, cats, and stoats. These new predators killed and ate birds, birds' eggs and their food. The moas and many other birds were consequently quickly driven to extinction.

Among New Zealand's unique native birds is a giant, bright blue-green, red-beaked rail called the takahe. It was once believed extinct, but in 1948 a few survivors were found in the Murchison Mountains on the South Island and, since then, a rescue operation has aimed to avoid losing this amazing bird for good. Takahes have a specialized diet comprising fern roots and the young shoots of some native grasses, which was a relatively secure food source, until red deer from Europe were released in the nineteenth century. Another threat came from weasels, which were released to control rats, but, like the foxes liberated in Australia to eat rabbits, didn't discriminate between pests and endangered species. Adept at finding takahe nests, the weasels plundered eggs for food, pushing the takahe to the edge of extinction.

Efforts to exterminate deer and weasels have not succeeded, and the natural vegetation of New Zealand (and thus the food of the takahe) is under siege from more than 1,600 established alien plants. Some of the last few takahes have been removed to the offshore Maud Island, where no deer and weasels have yet reached, but the species remains at grave risk.

Another of New Zealand's amazing unique birds which has had to be evacuated to offshore islands is the kakapo. Sometimes called the owl parrot, the kakapo is the largest parrot in the world. It is also the only one that is both flightless and wholly nocturnal. This rather unlikely combination adds up to a highly specialized bird. Kakapos are very slow breeders, and many of the plants that they most like to eat only bear fruit every few years. Under normal conditions, however, with plenty of space and no predators, kakapos thrived. For these unusual parrots, ecological disaster struck with the arrival of introduced opossums and deer that ate their favorite food plants. Cats and rats took the kakapos' eggs and young and, by the 1970s, they were virtually extinct. A tiny population survives, but these amazing creatures teeter on the brink of extinction and could soon follow the legions of other unique life forms driven into the abyss through the effects of introduced alien species.

The spread of non-native animals and plants around the world is breaking down the ecological distinctiveness of different regions and ecosystems that has been developed over hundreds of millions of years. Whether deliberately or accidentally introduced, these animals and plants have caused widespread damage to biodiversity, especially to highly specialized species that have been unable to withstand the ecological onslaught of aggressive, fast-breeding generalists. Neither have many creatures or plants been able to cope with the pressures placed on them or their exploitation by humans.

The kakapo is a ground parrot that has evolved a virtually flightless lifestyle and thus is unable to escape introduced predators.

hunted to destruction

On September 1, 1914, at the age of 29 years, Martha died. Her departure marked not only the end of her life, but the final demise of her entire kind—Martha was the last of the passenger pigeons.

A hundred years earlier, the extinction of this species seemed impossible, for the passenger pigeon was one of the most abundant birds ever to have existed. It is estimated that there were between three and five billion passenger pigeons at the time of European settlement of North America, with this single species believed to have made up 25 to 40 percent of the total bird population of what is now the United States. Early writers described flocks of passenger pigeons darkening the sky; single flocks could be more than a mile across and take several hours to pass overhead.

The birds migrated from Canada—central Ontario, Quebec, and Nova Scotia—south to the uplands of Texas, Louisiana, Alabama, Georgia and Florida. Their main nesting range was in the region of the Great Lakes and east to New York. A single nesting colony could cover many thousands of acres, with hundreds of nests in a single tree. One colony in Wisconsin was reported to cover 850 square miles with an estimated 136 million nesting birds. Amazingly, however, by the early 1900s, no wild passenger pigeons could be found.

Because passenger pigeons congregated in such huge numbers, they needed large blocks of forest for successful breeding. The early spread of farming led to the clearing and fragmentation of the great temperate forests of the northeastern USA. Although the birds were displaced, they still continued to breed. But, as their forest food supply decreased, the birds began feeding in fields in huge numbers and caused serious crop damage. Many were shot by farmers; many were taken for meat. Even this more systematic persecution didn't seem to make a serious impact on numbers, however, and it was only when professional hunters joined in that things started to change.

At first, the pigeons were shot out of necessity, for food. By the

early 1800s, this search for food had led to mass slaughter: birds were shot and netted at the nesting colonies; the chicks were knocked out of nests with poles. They were killed in their millions, and either eaten locally or shipped to distant markets—their abundance meant that they could be bought for as little as fifty cents a dozen.

By 1850 the destruction was in full force. By 1860 the birds appeared to get scarcer, but the massacre continued. One of the last large communal nesting areas was at Petoskey, Michigan, in 1878. Although the species was obviously in sharp decline, 50,000 birds per day were still killed there, and killing at this rate continued for nearly five months. Adult birds that initially survived

A large populations doesn't guarantee survival. The Passenger Pigeon was one of the most widespread birds in North America in the 17th century; it was extinct on the September 1, 1914. These three are stuffed examples in the Smithsonian Institute, Washington.

the massacre attempted to nest elsewhere, but were soon located and killed by professional hunters.

Laws were finally introduced to make it illegal to kill the birds within two miles of the remaining nesting colonies, but they were poorly enforced and few arrests were made until, by the early 1890s, the passenger pigeon had almost disappeared. In 1897 a bill was put before the Michigan legislature seeking a ten-year hunting ban, in the hope that the species might recover. But it was too late. The few surviving flocks and individuals couldn't muster sufficient numbers to form a colony and the last birds quickly died out without breeding.

The fate of this bird is not unique. Indeed, the most sustained and consistent pressure placed on the natural world by humankind arises from the direct exploitation of animals and plants. Since long before civilizations were established, people have hunted wildlife, for example, for food and skins. That long period of exploitation has resulted in very significant consequences.

The impact of early human exploitation of wildlife can be seen in the loss of the large animals that once made up what is known as the Pleistocene megafauna. The exact effect of hunting pressure as the main cause of extinction (compared to changes to the climate, for example) varies between continents. It is, however, increasingly clear that, in prehistoric times, human hunting pressure played a major role in the disappearance of a large number of animals.

If shown on a map, the wave of extinctions that swept through many large creatures in prehistory broadly follows the patterns of human colonization: a number of large marsupials were lost from Australia and New Guinea up to 50,000 years ago, around 12,000 years ago in the Americas, about 1,500 years ago in Madagascar, and between 900 and 600 years ago in New Zealand.

A lot of the large animals that lived in Africa (and which are still there) survived apparently because humans evolved in close proximity to them, thus permitting the wildlife to become gradually accustomed to the ever more dire threat posed by groups of humans armed with deadly weapons, rather than suddenly needing to deal with the threat. This did not happen elsewhere. Humans appeared and the big animals were not, in an evolutionary sense, prepared. Extinction rates appear to have been most extreme in places where humans arrived as already skilled hunters.

Since those far-off days, there has been a steady stream of hunting-derived extinctions. And they are set to continue. Indeed, of the thousands of animals regarded as at some risk of extinction, 28 percent of listed bird species and 29 percent of mammals are considered to be threatened mainly by hunting. The quest for food, furs, hides, horns, and shells thus continues to place a great many species under serious pressure.

Perhaps the best-known example of how hunting can affect wildlife is the decimation of whale populations. Although humans have taken whales for food, oil, and bone for many centuries, the threat dramatically escalated during the twentieth century with the advent of modern whaling fleets armed with modern technology, including brutally efficient equipment for killing and processing the animals. The fleets competed with one another in tracking down large whales, and blue whales, and several other species of large whale, were reduced in number to the point where they were in danger of imminent extinction.

Although there are laws in place both nationally and internationally to protect rare and declining species from the most serious impacts of commercial exploitation, they are regularly violated. The ongoing threat posed to Africa's black rhinos is a case in point. On top of the widespread destruction of its thorn scrub habitat, this creature now lives on the very cusp of extinction, having been ruthlessly hunted for its horn, a material that is highly valued in various clandestine markets.

The black rhino has perhaps suffered the most aggressive deliberate assault that has been directed against any animal and, as a consequence, its numbers have been drastically reduced. Today, there are now only about 3,000 black rhinos in Africa, a pitiful number compared to the 65,000 to 100,000 that roamed the continent in the 1960s.

Hunting has been an important factor in the decline of other rhino species too. The majestic prehistoric-looking Indian rhinos were not only hunted as prizes for horn, they were also killed as agricultural pests in tea plantations. Although by the early 1900s the population was so depleted that hunting was prohibited in Assam, Bengal, and Myanmar (Burma), these animals have not been made much safer.

Rhino horn is prized in traditional Asian medicine for the treatment of ailments including fevers and epilepsy. Asian rhino horn is believed to be more effective than African horn, and, despite protection, horn from these rare rhinos is still extensively traded. The 2005 wild population was estimated at 2,400 animals. This is a tiny population and these are slow breeding animals, but even they were increasing painfully from an even lower point, until their rise was set back by hunting. Between 1986 and 1995, about 450 rhinos were poached in India and about 50 in Nepal. In Manas National Park in India it has been estimated that hunting has recently killed more than 95 percent of the rhino population.

The illegal trade in protected wildlife is so huge that it has been estimated as the third-largest illicit market in the world after drugs and weapons. A less dramatic, or a less immediately obvious, threat, however, is that posed to biodiversity by different kinds of pollution.

The black rhino has been hunted to close to extinction mainly for its horn. The horn is highly prized as an aphrodisiac, though there is no scientific support for this claim.

pollution pressure

Since the start of the industrial revolution more and more chemicals have been pressed into use: in industry; for domestic purposes; and in farming, for example. Undoubtedly, the use of chemicals has improved human welfare immeasurably, and helped to accelerate the process of development. There have, however, been a series of side-effects that have had a serious impact on wildlife populations.

pesticides

One of the first very widespread consequences of chemical use to come to public attention was in relation to pesticides. In the years following the Second World War, the use of farm chemicals exploded. They were used to control various kinds of agricultural pests, and it seemed that the range of new miracle applications could thereby boost yields, increase farmers' incomes, and consequently make food cheaper for consumers. The potential for increasing food production in countries with rapidly expanding populations seemed vast. But, as time went by, it became clear that there was a high cost.

Significant fish and bird kills have resulted from the legal application of pesticides, such as the 1991 death of more than one million fish in Louisiana. The use of powerful pesticides on banana plantations in the tropics has also killed huge numbers of fish in nearby rivers and streams. Some of the effects are, however, more subtle than simply causing the outright death of large and obvious species. One result has been changes to the relative populations of insects; another has been alterations in the number of soil invertebrates and microorganisms; there have also been changes to the plant communities found in particular farmed landscapes, or adjacent to them. These impacts have in turn affected the overall health of some ecosystems, and thus influenced nearly all of the wildlife in the area, even if the pesticides used are not directly toxic to all species.

Long-term effects can result from changes to the food chain. For example, when pesticides reduce the insect populations that comprise the diet of birds, causing food shortages for chicks during the breeding season, this leads to dramatic population

reductions. Fish that feed on aquatic insects may also show stunted growth in areas of heavy insecticide use, because their food sources are diminished.

Pesticides include a wide range of chemicals. Some are designed to kill insects, for example, by attacking their central nervous systems; others are manufactured to target plants; others kill microorganisms, including viruses and fungi. It is not only in the farmed landscape that pesticides are used, however. Today many are also used in urban areas and in gardens.

In some countries pesticide residues are very widespread. One study in the United States found that 100 percent of all surface waters sampled, and 33 percent of major aquifers, contained one or more pesticides, and pesticides were identified as one of the major causes for the poor quality of many streams. In the USA pesticides have also been identified as a potential contributory cause for the decline in some amphibians and in some species of beneficial insects, including those which pollinate crops.

Some chemicals are highly stable and can, therefore, persist in the environment for long periods. The pesticide DDT is one such substance. For decades it was successfully applied in large quantities to protect crops, but it did not immediately break down, and traveled through food webs. DDT thus built up in the bodies of animals, accumulating in particularly high concentrations in top predators, especially in birds of prey, such as the bald eagle. This prevented them from breeding, in some cases because their egg shells became too thin, and they broke when the birds incubated.

The long term effect of some chemicals is often complicated. DDT caused a crash in the population of birds of prey because at certain levels it interfered with the calcium metabolism involved in the production of eggs. The resulting eggs, such as these peregrine eggs, had very thin shells, which were easily broken by the adults when incubating. The effect on the population was devastating. The population crashed in the 1960s and 70s, only recovering after DDT had been banned for more than ten years.

nitrogen

Artificial farm fertilizers have also had a major impact on biodiversity. Nitrogen is an important plant nutrient. It occurs naturally and, indeed, comprises nearly four-fifths of the Earth's atmosphere. In this gaseous form it is inert, however, and has little direct impact on the natural world. When "fixed" though—this occurs through natural processes, such as by the action of bacteria like those living in the root nodules of various members of the bean and pea family—it accelerates plant growth.

At the start of the industrial revolution, the quantity of fixed nitrogen rapidly increased when human sources were added to natural sources. This occurred firstly because of incidental release from coal burning; then through the manufacture of artificial nitrogen fertilizer for use in agriculture.

More recently we have used a lot of manufactured nitrogen fertilizers. One recent study estimates that more than half of all the synthetic nitrogen fertilizer ever used has been used since 1985, and this rate continues to increase. Though used to stimulate crop growth and increase yields, a lot of the fertilizer hasn't stayed in the fields and there is now an unnaturally high level of biologically active nitrogen cycling in the atmosphere, on land, and in lakes, in rivers, and in the sea. Indeed, fixed nitrogen levels are now at about double the original natural level. This is having profound effects, causing farming-related nutrient pollution problems around the world.

These environmental changes include a process called eutrophication. This occurs when nitrogen, and other nutrients, fertilize a water body and stimulate a "bloom" of microscopic algae. The rapid algal growth in turn reduces oxygen levels in the water to the point where animals, such as fish, die. One projection suggests that the continued intensification of farming will lead to more than a further doubling in nitrogen-driven eutrophication of ecosystems. Nitrogen pollution can also cause acidification of freshwater and terrestrial ecosystems and can change the nature of habitats, causing some species to be lost as more aggressive plants survive and thrive on the extra fertilizers.

Unnaturally high levels of nitrogen not only occur near to the

farmland where it is used, but can travel long distances in the environment. One study estimates that many of the world's most important biodiversity "hotspots" will, in the coming decades, be deluged with excess nitrogen. This is expected to destroy some species, causing ecosystems to become more simple.

In severe cases nitrogen pollution can lead to the creation of so-called "dead zones" in the sea. About 150 such areas have already been identified: some are quite small, covering just a few square miles; others are much bigger, embracing some 27,000 square miles of sea. One of the best-known "dead zones" is in the Gulf of Mexico, where vast inputs of nutrients arriving via rivers (such as the Mississippi) and draining some of the most intensively farmed land in the world, has led to a vast area of depleted oxygen.

Increased levels of nitrogen result in dramatic blooms of algae. This drastically reduces the amount of sunlight that permeates the water and effectively kills most other plant life in the water, which is a classic symptom of nitrogen enrichment.

industry

Pollution arising from industrial sources has also had a serious impact on wildlife, including highly persistent substances that have interfered with animals' biological systems. PCBs were produced for electrical insulation purposes, but, as with the DDT used to control crop pests, the chemicals traveled into the environment to accumulate in the bodies of top predators far away from where they were produced. Although manufactured in industrialized countries, PCBs have ultimately been found right around the planet; for example, they have been discovered in high concentrations in the bodies of polar bears, and have caused damage to the bears' hormonal systems to the point of causing sex changes.

Agricultural and industrial pollution can be an especially important threat when they affect corals, because most corals mass spawn and produce floating eggs and sperm. If pollutants impact on the breeding ability of corals, reefs over a large area can be endangered.

Another source of pollution comes from power stations. Sulfur and nitrogen compounds released from coal-fired power stations in particular have led to so called "acid rain." This has caused forests to die back dramatically across large parts of central Europe and led to the ecological demise of thousands of lakes in Scandinavia. Insects, fish, and sensitive plants have been directly affected, causing widespread change to ecosystems. Acid deposition from power generation remains a major threat to biodiversity in many parts of the world.

The effects of pollution released into the environment are many and widespread. While some ecological changes have been closely documented, others remain as yet poorly researched and understood. What we do know, however, is that the combined impacts of pollution, habitat destruction, introduced species, and the direct exploitation of wildlife is leading to a mass extinction of species and populations on a scale not witnessed on Earth for millions of years—possibly even tens of millions of years.

"acid rain" effects are often most visible downwind of heavy industry.

so what?

Should we be bothered at what is happening to the natural world? After all, the huge improvements in the material well-being of many people surely vindicates the manner in which we have used nature. For thousands of years people have taken what they needed from their environment; in many respects that, in itself, can be seen as part of a natural order, part of the process of evolution—as one species changes, it affects others as it seeks to improve its security and comfort. It is not so simple, however. The scale of change now taking place because of our activities could mean that we pay a huge price ourselves. It is not so much a question of why should we conserve the Earth's living resources, therefore; it is more a matter of how can we do it so as to ensure our own well-being, even survival.

Kamut wheat is the ancient ancestor of modern durum wheat and could be important in developing other types of cereal in the future.

food and raw materials

Although it is sometimes not immediately obvious, nature provides immense riches and benefits for humankind. Many of these are irreplaceable. For a start, everything that we eat is derived from once-wild species—wheat, chickens, pigs, apples, tomatoes, and everything else that we require for our dietary welfare is provided by species that have wild ancestors. Given the tiny handful of species that we use for food (compared with the vast range of biodiversity that, at present, we do not) it seems most unlikely that we have identified all of the potential sources of food that might exist in the ecosystems that we are currently wrecking with such abandon.

Humankind faces several very considerable and interlocking challenges in the decades ahead—coping with rapid population growth, climate change, and the continuing challenge of ending poverty— and it would be prudent for us to preserve as much choice as we possibly can, not least in respect of the potential food species that we might be able to draw on in the future.

This is not simply a case of being able to find new food species that might help us adapt to future conditions, it is also—crucially— a matter of maintaining the wild relatives of the species that we have already domesticated. The vast wealth of genetic diversity

that remains in the wild relatives of the plants and animals that we already use is of huge importance. One estimate suggests that about half the annual increase in crop production comes from the incorporation of new genes from populations of wild relatives. This input of new genetic material helps the crops achieve enhanced resistance to pests, disease, and drought, for example. We may need this boost to farm production even more in the future.

Natural ecosystems are also sources of great quantities of raw materials. Carefully managed semi-natural forests can provide timber indefinitely, while also conserving much of their original biodiversity; mangroves and coral reefs act as the nursery grounds for a significant proportion of the fish that the world consumes.

science

Another direct benefit that we rarely acknowledge is the role played by biodiversity in medicine. For example, more than 2,000 tropical forest plants have been identified as having some form of anticancer properties. Although the potential for improving human health using the services provided by biodiversity is clearly vast, we are destroying that potential before we have barely begun to understand it. Humans have only tested one in ten tropical forest plants for their anticancer properties so far, and only intensively screened about one percent of them. Many are going extinct before we have even assigned them names, never mind assessed their potential to help us.

Tropical rain-forest plants have proved an especially prolific source (although by no means the only one) of plants with medicinal values. This is, in part, because of the intense ecological competition that takes place in these ecosystems, for example, between herbivores and plants. The equivalent of chemical-weapons races have broken out under these circumstances, leading to plants manufacturing a wide range of compounds as a means to promote their chances of survival. Many of these compounds can help us to survive too.

The medical challenges that have already been addressed with rain-forest species are varied. Malaria, widespread and mosquito-borne, has for decades been treated with quinine, first found in the bark of the cinchona tree. The muscle relaxant curare, which is used during surgery, was first extracted from a species of a vine. It was used for centuries by indigenous peoples to poison arrows and darts. Tropical rain-forest animals have also made contributions to medicine: a species of Amazonian frog contributed a compound that helps in the treatment of strokes, seizures, depression, and Alzheimer's disease.

The services provided by the Earth's living systems go much further than what they can do directly to help our species, however. The very functioning of ecosystems depends on biodiversity—to recycle nutrients, dispose of waste and to purify water, for example. Biodiversity also has a central role in maintaining global climatic stability: the tropical rain forests act as a global air conditioner, absorbing and storing carbon dioxide from the air and releasing oxygen. These systems also play a major part in regional water cycles, generating rain and sustaining and moderating the flow of rivers.

There are also scientific reasons as to why we should invest effort now in saving the Earth's biodiversity. This is the only planet that we know sustains life, in its vast variety, bound together in millions of subtle interactions. In losing that vast reservoir of diversity, we are allowing an utterly irreplaceable asset base to be removed. And it cannot be brought back, certainly not in its present form. And, even if life were ever to be restored back to its former diversity, the lesson from the fossils is that it would take millions of years.

tourism and the fight against poverty

Wild ecosystems are also an increasingly important asset for tourism. This sector is the number-one industry in many countries and the fastest-growing economic sector in terms of foreign exchange earnings and job creation. International tourism is the world's largest export earner and an important factor in the balance of payments of many countries. A high proportion of this economic activity is driven by people wishing to spend time with nature: watching birds, trekking through wilderness, and diving in reefs, among other activities. The destruction of these assets is very likely to lead to reduced future revenues in a vital sector.

Despite all these practical and very immediate reasons as to why we should invest time, effort, and resources into conserving biodiversity, it is still widely believed that there is a choice between conservation and development, and that it is necessary to sacrifice biodiversity in order to improve human well-being, especially for the world's poorest people.

The Millennium Ecosystem Assessment, published in 2005, rectified this popular misconception, however. This very thorough and comprehensive survey of the state of the natural world concluded that, if we do not take more active steps to protect natural systems and biodiversity, poverty will get worse, not better.

The assessment concluded that the destruction of the natural world was an increasingly significant barrier to the achievement of poverty-reduction targets. It said that United Nations' goals to halve poverty and hunger by 2015 will not be met, and that hunger and malnutrition will remain problems even in 2050, unless governments and societies pay greater attention to what nature does for humanity, and ensure nature keeps on doing it. This is not least because poorer societies depend directly on the natural world for their well-being: on productive soils, reliable water supplies, prolific fisheries, and wild meat and timber from natural forests, among other things. As "development" pressures remove or degrade these services and assets through unsustainable patterns of land use and resource management, so the prospects for human development decline rather than improve.

Other less tangible, but very real, values that nature provides for us must also inform our approach. Our spiritual well-being is certainly served by nature. Not only is this linked to our direct experience of wildness and wildlife, and the restorative properties

that such experiences can bring, nature also inspires art. From music to writing and from painting to sculpture, our humanity is fundamentally informed by our esthetic relations with the nature that gives us breath, and which sustains us all—even in our ever-more-detached urban environments.

our moral responsibility

There is, however, above and beyond all this, a more profound and compelling reason why we should conserve the Earth's natural diversity: we have a moral responsibility. The Earth's great natural riches have emerged following three and a half billion years of evolutionary change. Does any single species have the right to destroy much, or even most, of that unique planetary heritage?

Our present actions seem all the more indefensible given that what we are now doing is with our full knowledge. We know the impacts of habitat loss, of introduced species, of hunting, and pollution. Many argue that we have no right to destroy what we did not bring into being but, even from a purely humancentric perspective, there is a strong moral case to change our relationships with nature. Having taken what we need to maintain our welfare, is it right that people today should knowingly deny future generations the opportunities that we and our immediate ancestors have enjoyed? Looking back to today from the near future, it is hard to see how our children will thank us for allowing—let alone that we consciously know that we are doing it—countless species to be committed to oblivion.

Fortunately, all is not yet lost. It is possible for us to take steps now that will enable us to pass on intact much of the natural world and the irreplaceable services it provides for us.

Turtles are not worried by the sound of the planes, but they are disturbed by the hotels that are built to accommodate the tourists who are attracted by the same beaches that the turtles need to breed on.

no time to lose

Every now and again it seems that humankind is offered the opportunity to turn back the clock of extinction, a second chance placed before us, and at times we rise to it.

The rediscovery of the takahe and kakapo in New Zealand gives us the chance to rescue what we had previously assumed were already lost species. Emergency programs now in progress could yet save these unique creatures. If successful, these last-ditch measures could pass on an irreplaceable gift to the world: the continuing presence of animals that have their origins in the ancient isolation, and ultimately in the far-flung ancestry, of our planet's long history; irreplaceable creatures, brought back from the abyss. And with time, if we are lucky and resourceful, restored to their former status as fully wild creatures inhabiting their native lands.

In 2004 scientists made what was thought to be the most amazing rediscovery of all of an already "extinct" species. The ivory-billed woodpecker was a gorgeous creature of the lowland forest of the southeastern USA and Cuba. In the land of the world's biggest economy, where large-scale destruction, fragmentation, and degradation of habitats has been conducted across the continent, a bird that has been assumed to be lost since 1935, was found in Arkansas: alive. The discovery of this large handsome red, black and white bird was all the more amazing because of the large number of bird-watchers in that country who hadn't noticed that a few of these birds remained. More than 60 years had elapsed since the species was last seen in the USA (although birds hung on longer in Cuba), until what was thought to be an ivory-billed woodpecker was found in a patch of undisturbed forest.

Others have definitely recently come back from the grave. The Cebu flowerpecker is another bird thought extinct for decades only to be rediscovered, albeit in a grave state, in dense forests on its island home in the Philippines. Jerdon's courser went missing for almost a century before it turned up in scrub forest in Andhra Pradesh in India. There may be many others.

Although we cannot rely on these fortuitous rediscoveries as the basis for a conservation strategy, we should be encouraged by them, and take confidence that there is still a great deal that can be achieved. Even at this late stage, with a significant mass-extinction episode already underway, we can still take action to preserve, protect, nurture, and sustain the biodiversity that is at the very heart of the uniqueness of our world.

We need to be clear, however: about priorities; about the measures needed; and who will be responsible for what action and by when. We also need to mobilize the relatively modest resources needed to make the critical difference, and to do it in time. That means taking large-scale and broad action immediately. Our plan needs to be orchestrated, and to work simultaneously locally, nationally, and internationally. It is complicated, but it is possible.

What was thought until recently to be one of the last Ivory-billed Woodpeckers—a juvenile. In 2004 adults were seen in the same area as this picture was taken. We have one last chance to help this species survive.

nature's place

The essential measure we need to take is to secure the most important areas of remaining natural habitat. It is now widely understood that habitat destruction and ecosystem degradation and fragmentation are the main causes for the present rapid rate of biodiversity loss. Protecting and restoring habitats is thus in many cases the critical conservation strategy that must be pursued.

protected areas

Most countries now have some form of protected area system, for example, a network of national parks, and in recent years the number of such reserves has dramatically increased. There are now more than 100,000 protected areas worldwide, covering nearly 7.3 million square miles: that is 12 percent of the global land area, an area about equivalent to the size of South America. This is certainly an improvement compared to even the recent past—it represents a ten-fold increase on protected areas in the early 1960s. Europe has the most individual protected sites—43,000— but Central and South America have the highest percentage of land protected for conservation purposes, with more than one-quarter of their territory set aside.

The United Nations estimates that between 10 and 30 percent of the planet's vital natural features, such as the Amazon rain forests, tropical savannah grasslands and the Arctic tundra are now officially designated for conservation purposes. Some of these reserves are huge, such as the Great Barrier Reef Marine Park in Australia. This vast natural feature, visible from space, is conserved in a marine park covering more than 128 million square miles. The Amazonia forest reserve in Colombia is only slightly smaller. Others, by contrast, are tiny, covering patches of remnant habitat, including some that are now surrounded by urban development.

The coverage is not only uneven as far as the size and distribution of reserves is concerned, but also in relation to the kinds of habitat protected. Some major habitat types are under-represented—for instance, fewer than 10 percent of the world's major lakes are included. In the marine environment the situation is far worse, with less than half of one percent within protected areas.

how should we choose

Most protected habitats have been carefully chosen so as to maximize conservation benefits. Research by ecologists has enabled us to better understand which parts of our planet are most important for biodiversity conservation. This gives a better idea of where protected areas can be most beneficially allocated. One approach in this respect has been to identify so-called "hotspots," which are areas with relatively more biodiversity than others.

Another approach is to identify those places where wildlife is most unique or most threatened. Very often these places coincide: where biodiversity is very rich, unique species and habitats often face a high risk of extinction. Many areas of unique forest fall into this category, for example, several of those in East Africa, and in the subtropical and temperate zones of the Andes. Madagascar accounts for 2 percent of Africa's land area but this incredible island has 10,000 species of plants, 80 percent of which are found nowhere else in the world. Deforestation on the island is in many parts nearly complete, placing a high proportion of whole families of animals, such as the lemurs, at risk. Incredibly, in 2006 alone, some nineteen new species of these primates were described for only the first time, perhaps just in the nick of time.

Worryingly, many threatened species do not occur within any protected area, meaning that, according to one recent study, hundreds of endangered species are vulnerable: 260 kinds of mammals, 825 amphibians, and 223 bird species. With this information researchers concluded that a major expansion of protected areas was needed in order to avoid a wave of extinctions in the near future; unsurprisingly, perhaps, it was areas of tropical rain forest and islands that were most in need of urgent protection. This analysis further estimated that, by protecting a further 2.6 percent of the world's land area, it would be possible to include about two-thirds of those endangered species which, at present, have no protected habitat.

But levels of species richness and the location of threatened species are not the only way of identifying where protected areas should be designated. Another way to approach this challenge is to understand which areas are most distinct for animals and plants in

terms of where unique branches of evolution are found. Researchers recently took a close look at the undisputed botanical hotspot that exists around the Western Cape of South Africa. The thousands of species that occur in this region were then fed into an "evolutionary tree." Through a careful analysis of how species, in an evolutionary sense, related to one another, scientists put together a full relationship map of the Western Cape flora. This was the largest such exercise ever carried out, covering more than 9,000 staggeringly diverse species, many of which are at risk of extinction and only found in the Western Cape.

The mapping revealed how all of the plants in the Western Cape are closely related to one another. A similar mapping of the Eastern Cape found a different picture: although there are fewer species of unique plant found there, the ones that are present are less related to each other than in the Western Cape. Rather than dense clusters of twigs at the end of fewer evolutionary branches, as seems to be the case in the Western Cape, the eastern plants represent more branches.

Madagascar is referred to as a centre of endemism because it is one of the richest areas in the world for unique species, most notably the lemurs.

The Cape Sugarbird has evolved to pollinate proteas, which are only found in the Western Cape. Without the Sugarbird the proteas would be unable to pollinate, without the proteas the Sugarbird would have no plants to feed on

While more traditional conservation strategies have tended to focus on these most diverse areas, it is also be important to understand the level of difference between species in any given region. Evolutionary novelty—a term that describes how unique particular life forms are—is now increasingly recognized as an important element of biodiversity that needs to be reflected in the placement of protected areas and other conservation measures.

It is also vital that conservation strategies, including protected areas, focus on maintaining the diversity within species. In this sense, it is vital to conserve the diversity of different populations, not just part what is reflected in any one species. This is especially true if we wish to maintain the potential usefulness of biodiversity in meeting human needs in perpetuity.

Whichever way protected areas are identified, it is vital that they are then managed in the correct way. Some areas with high levels of biodiversity contain a rich mixture of habitats, including various levels of forest regrowth, for example, thus providing different-aged trees that may support different communities of wildlife.

Habitat diversity can be increased through various kinds of human use. Some need to be strict nature reserves, where access is tightly controlled and different economic activities such as logging and hunting are forbidden; some important habitats actually require human activities to conserve their wildlife, such as a certain level of grazing, some selective forestry activities or water-level management. These interventions tend to be more important in areas where only small fragments of natural habitat remain, and where it is necessary for human activities to create conditions that would only occur naturally in far larger patches of natural habitat. This is especially the case in heavily settled parts of Europe.

Even if rare animals and plants remain in well managed habitats, if these are scattered in isolated locations, each population is more vulnerable to local (and, ultimately, total) extinction. This is one reason why it is necessary in some parts of the world to not only protect what remains but to recreate, restore, and expand habitats that have already been lost, fragmented, or degraded. If this is done carefully then it will be possible to retain far more biodiversity than would otherwise be the case.

restoring the earth

Right across the planet, habitat restoration and recreation programs are underway. Many are, as yet, small-scale and geographically limited initiatives, but the fact that such programs are now expanding marks a historic turning point. No longer is conservation simply a question of avoiding damage, or conserving the high-quality natural areas that remain (often just fragments of once-vast ecosystems) it is now about putting back what has been lost.

This shift in emphasis is already paying dividends. From the degraded fenlands of eastern England to the Salt River near Phoenix, Arizona, from the heavily degraded and polluted Chesapeake Bay on the East Coast to the fragmented Atlantic rain forest of eastern Brazil, conservation groups, official agencies, and volunteers are seeking to turn the tide of habitat loss into a positive agenda of restitution. In some places it is already working, and not only for biodiversity.

Many other social and economic benefits can be derived from carefully planned restoration programs. These include the protection of water supplies, reducing the risk of flooding, and increasing resource supplies, such as timber. Promoting economic and conservation benefits in tandem can clearly be a hugely powerful positive force for securing the future of life on Earth.

people as partners

Protected areas and habitat restoration can work especially well when local people work as partners in realizing conservation aims. While, for much of the twentieth century, conservation efforts focused on the protection of what were regarded as pristine and uninhabited or sparsely populated wildernesses, it is ever clearer that people have a central role to play in achieving conservation goals. People have often been seen as intruders into nature, an automatically negative force undermining biodiversity conservation. This view is quite wrong, however.

Humans (or at least human ancestors) have lived in Africa for millions of years, in Asia and Europe for several hundred thousand, in Australia for fifty thousand years, and in the Americas for at least the last twelve thousand years. Certainly, early humans took their

toll on species, for example hunting to extinction many large animals in prehistoric times. The impact of people has by no means been wholly harmful, however: far from it.

Indeed, many habitats that today are regarded as "natural" have, for long periods, been shaped by the activities of humans. Before 1500 and the arrival of European colonists, there were an estimated six million native people living in Brazilian Amazonia. These people were not passive inhabitants of this complex ecosystem. Over millennia, these indigenous people had caused subtle but major impacts on the apparently untouched natural forests the settlers found when they arrived.

The native forest people selectively burned small patches of land to grow crops, such as manioc and sweet potatoes. When these primary crops were harvested, fruit trees were planted, not only providing a further crop of food but also attracting fruit-eating animals. The pattern was repeated in other areas, providing

The Sonoran Desert is one area where human influence is important because, here, local farmers actually increase the biodiversity.

Indigenous people usually live in a delicate balance with their surroundings, for example, only grazing as many animals as the area can sustain indefinitely. If that is fewer than will sustain them, they move from area to area as the grazing runs out.

secure food supplies, while at the same time sustaining the forest. Over long periods this shifting cultivation subtly changed the ecosystem in ways that were often beneficial to wildlife. The native forest inhabitants of Amazonia nurtured their environment far more thoughtfully than the loggers, miners, and ranchers who have largely replaced them.

Comparable situations have arisen in other areas. It is believed that shifting cultivation may be essential to the welfare of some of the large endangered mammals found in the rain forests of Indochina, such as koupreys (a kind of rain-forest cattle), elephants, and tigers. In the Sonoran Desert that straddles the Mexico–U.S. border, it seems that the traditional land use practices of local farmers increased the biodiversity in the oases that are

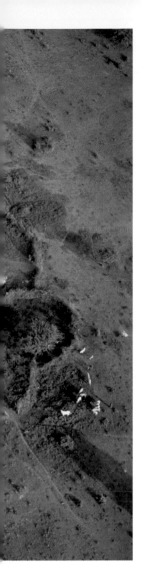

found there. In one protected area from which people were excluded, biodiversity actually declined.

Local human inhabitants of complex systems, using traditional land use, are the most important resource available in saving the planet's biodiversity. But, like the wildlife, indigenous and other communities who live close to nature are being wiped out. By the year 2000 there were fewer than 250,000 native people still living in the Amazon basin. This decimation had various causes: they were deliberately killed for their land; they were pressed into slavery; and they died from introduced diseases to which they had no resistance. The genocide has certainly caused changes to the forest, not only by removing one of the most important ecological agents, but also in destroying the forest's natural custodians.

Native peoples and other communities living close to nature can be powerful allies in sustaining the Earth's natural riches. Finding common cause with local people can thus be a very powerful and effective conservation strategy. Reviewing the present-day map of rain forests in the Amazon basin amply demonstrates the point: in many parts the areas of intact forest coincide very closely with the lands still controlled by the traditional indigenous inhabitants. Their culture and lifestyles are far more conducive to sustaining the forests than the extractive depredations adopted by the recent incomers seeking the lucrative financial returns that can accompany the ravaging of natural resources.

In this way conserving biodiversity is bound up with the land rights of indigenous people, and thus questions of social justice. If social justice and land rights can be secured for the world's indigenous peoples and other societies living sustainably in and near the most important natural systems, then it will be possible to dramatically improve the long-term prospects for large areas of biologically rich ecosystems, as well as protecting the Earth's incredible and equally irreplaceable human cultural diversity. Supporting local people in maintaining their traditional lands and lifestyles (whether in protected areas or not) must be a central element in efforts to save life on Earth.

cutting pollution

While a combination of protected areas, ecosystem restoration, and sustained traditional livelihoods and cultures can transform the fortunes of biodiversity, wider action is also needed. Factors outside these natural systems also play a vital role. For example, controlling excessive fishing in coral reefs is a vital measure for conserving marine biodiversity, but this will often on its own not be sufficient. To maintain the full integrity and diversity of reefs it is essential that different kinds of pollution are also minimized.

In many areas, development programs taking place at a distance from important marine habitats also need to accommodate reef conservation in their design, for example, by ensuring that new tourist hotels install modern sewage-treatment facilities.

Some agricultural activities also need to be changed to limit pesticide and fertilizer contamination. Agricultural policy is a complex area, however, which links into questions of food security, international trade agreements, and national subsidy regimes, in addition to elaborate pesticide-approval rules. History, though, shows that even in this difficult area improvements can be made.

Following some initial controversy regarding the link between DDT and the disappearance of predatory birds, this chemical was banned from farm use in many countries. The recovery of severely depleted bird populations followed soon after. There are still obviously many impacts arising from excessive farm-chemical use that must be addressed, however.

Many of the most biologically harmful sources of industrial pollution have also been cleaned up in some countries. Regulations and policies requiring industry to either minimize or end the use of certain harmful substances have led to less pollution and, therefore reduced pressure on wildlife populations. Again, more needs to be done, especially in rapidly industrializing nations.

The impact of acid rain has also been reduced in many countries through laws and policies that have required the pollution from coal-fired power stations to be drastically reduced. In parts of Europe some ecosystems have thus begun a long process of recovery. In many parts of the developing world, similar action is now needed, in part through technological developments driven forward by legal controls.

setting aside land in areas of intensive farming creates **wildlife enclaves**, which benefit natural ecosystems.

the sperm whale lives thousands of miles from the site of DDT or PCB use, but all sperm whales have traces of such chemicals in their fat.

protecting the rarest

Although the passenger pigeon did not survive into the age of more rational wildlife management, the timely establishment of new laws and controls on their exploitation have worked in helping other rare and declining species beat the grim reaper of extinction.

There is now an abundaance of national legal controls across the world enacted so as to protect animals and plants from over-exploitation and persecution. These controls range from outright bans on the hunting, harvesting, or collection of certain species, through to rules that, while designed to sustain populations, permit some exploitation. For example, closed seasons on hunting are one means of doing that.

There are also a number of international rules and laws that have been agreed. Perhaps the most important global measure is the Convention on the International Trade in Endangered Species (CITES). This treaty, now signed and ratified by nearly all countries, lists which animals and plants are banned from international trade and sets out which others may still be traded so long as there is no threat to the survival of the species. There are problems with this agreement. For example, a lack of available resources in some countries prevents them from enforcing CITES effectively, but it has undoubtedly saved many species from being driven to extinction.

With the great whales being hunted close to annihilation, various nations met throughout the 1930s in an attempt to bring order to the whaling industry. An international agreement aimed at more effectively managing the hunting of whales was finally signed in 1946. The agreement of the International Convention for the Regulation of Whaling in turn led to the establishment of the International Whaling Commission, a body that would promote scientific research on whale populations and act as a forum at which governments would, using the basis of that research, agree what to do. With too little effective management of whales under this treaty, however, the slaughter continued.

Blue whales were nearly wiped out, and many other species were successively targeted by the whaling fleets. Catching blue whales was finally banned in 1966, although 99 percent of their original population had been killed. In 1986 a complete moratorium on commercial whaling was introduced. Fortunately, measures taken to protect whales have now had some positive effect—some species are beginning to increase in number—although many species of whale are still classified as in danger of extinction. Controversy has nonetheless raged over the status of different whale populations, and in 2006 commercial whaling was, on a limited basis, restarted.

But laws and treaties are only as good as their enforcement. The lucrative financial rewards that can come from illegal hunting and collecting tempt many to break the law meaning that many species remain at risk even though they are, in theory, protected.

One protected, yet critically endangered, species threatened with total extermination through illegal hunting is the Cambodian crocodile. A tiny remnant population of many of the remaining crocodiles was recently found living in the Veal Veng Marsh near to a local community who, fortunately, revere and even pray for these rare animals. However, these creatures are still highly sought after by crocodile poachers throughout Asia. An adult crocodile fetches up to $2,000 on the black market, which is an absolute fortune in local terms, where the average financial income is just a few hundred dollars per year.

The same financial attraction continues to lead to the decimation of many other rare and protected species. A single rhino horn can fetch tens of thousands of dollars, a considerable prize even for the already wealthy. Some rare parrots, such as various macaws and cockatoos, can fetch handsome prices of up to thousands of dollars per bird. Perhaps it is little surprise then that an estimated one-third of the total trade in wildlife is illegal. With financial incentives like this, it is clear that strong enforcement backed by tough penalties must be a central aspect of strategies to save species from exploitation. So, too, are measures to educate buyers and to discourage the creation of markets in rare species in the first place.

breeding and release

When a species declines to the point at which it is nearly lost, or indeed has already become extinct in the wild, then captive breeding can be a useful means (or in some cases the only means) to aid its recovery.

The European bison, golden-lion tamarin (a small Brazilian monkey), and the black-footed ferret are among animals that have benefited from captive breeding. Some species that went extinct in the wild have thrived in zoos: Arabian oryx, Père David's deer, and Przewalski's horse are among them, with the latter species recently reintroduced into the wild in Mongolia. For some others, such as Spix's macaw, which have been wiped out in the wild, there is no option other than to promote well-organized captive breeding in the hope that, should this be successful, birds will be released back into their natural habitat one day.

Nearly one-fifth of all mammals and one-tenth of bird species have been successfully bred in captivity. If nothing else, breeding in collections and zoos can remove the need to catch more wild individuals to sustain captive populations, which, in their turn, can be vital in rallying public support for conservation programs. When it comes to garnering public backing, there is nothing more inspiring than living, breathing creatures. Sustaining some biodiversity in captivity can thus be an important part of a wider conservation strategy. Having seen a live orangutan, for example, more people would be inclined to back policy changes to combat deforestation and save the animal's habitat than if they had had no direct experience of such animals.

There are serious limitations to captive breeding, however, including the high cost. Another is the naturally limited potential, given the relatively few species that can be effectively bred and maintained at any one time, even if all zoos and collections were participating in planned and coordinated breeding programs. Questions have also been raised about ownership of animals, however, and there are issues surrounding getting effective participation in programs. In addition, outbreaks of disease can have disastrous consequences.

only one spix—macaw was left in the wild: it was effectively extinct. **captive breeding** has resulted in an increase in the population and birds may be reintroduced to the wild in the future.

Even with captive breeding and the control of aggressive invaders, the biggest threat to most species is the wholesale destruction of their habitat.

So, even if captive breeding is a useful or necessary part of a recovery strategy, then it is, of course, essential that natural habitat is preserved or recreated at the same time as animals are reared so that they can be subsequently released. Certainly, it is far more sensible to ensure that animals are protected in nature, and never need emergency breeding in zoos and other captive collections. It does happen, though, and captive breeding can be the last chance for some species whose decline and disappearance has not been stopped in other ways.

controlling aggressive invaders

Considering the widespread damage that has been caused by introduced species, it is clear that considerable conservation benefits can be derived from the control and removal of these invasive alien species. Preventing their arrival in the first place is better still. This is easier said that done, however.

The successful eradication of established invaders is rare, and measures to control unwelcome alien species vary enormously in their effectiveness. It seems that success is more often linked to long-term commitment and consistently following though measures, rather than to any particular method. Shooting, trapping, and the use of poisons can all play a role, depending on the circumstances and the type of invader.

Although less high profile than habitat protection measures, action to control invasive species is certainly a major conservation priority. Many governments, such as that of New Zealand, have recognized this and have taken action to control the spread of animals like rats and cats, to islands in particular, and have taken steps to limit the risk of inadvertent introductions. Where they have escaped and gained a foothold, in some cases they have been removed again.

Failure to be effective in controlling introduced species could lead to severe global consequences, not just that of impoverished ecosystems comprising aggressive cosmopolitan species. There are also serious implications for resources that are important for humans, including farm output, forestry, and the productivity of fisheries. All of these can too be affected by alien invaders.

People may not have known about the potentially devastating impact of introduced creatures centuries ago. Today, though, we know a great deal and should be making every effort we can to avoid the release of exotic species, as well as investing far more resources in removing the ones that have already become established to the detriment of native animals and plants.

high stakes

We are entering a world in which degraded and fragmented ecosystems are becoming the norm. Suffering from stress caused by pollution, and with ecological relationships simplified by the impact of aggressive aliens, habitats are becoming more ruined. Where large and valuable species have been removed, critical system links are being cut and natural habitats taken even closer to total collapse. It needn't be this way.

We still have time to avoid the loss of much of the remaining biodiversity, but we are approaching the last moment for many habitats, species, and populations. If we act in time we can make a decisive difference; if we carry on as we are now, then there are undoubtedly grave consequences in the near future for life on Earth.

Collaboration with local communities to achieve large-scale habitat conservation, backed up with pollution-control measures and properly enforced species protection, augmented where appropriate with captive breeding programs and the control of introduced animals and plants, could still make a massive difference. In fact, it already is making a difference. But it is not enough, however, not by a long way.

We are the custodians of life that took 3.5 billion years to emerge in its present unbelievable richness. The role of humanity as the steward species of the Earth is sometimes a difficult concept to grasp, but there are two reasons why we must regard the present situation in this way: the first is because humankind is now, without doubt, the most important ecological force on Earth; the second is that we know it, and we are increasingly aware of the consequences.

Contrary to some popular wisdom, the economic resources needed to effectively discharge this role of custodian is, in a global sense, quite manageable. For example, one calculation finds that the purchase price for the entire Amazon basin rain forest (were it for sale) is equivalent to the total world military expenditure for only about three weeks. Although a hypothetical example (it may be much better to grant local people land rights rather than to purchase forest areas), this broad financial comparison illustrates how it is actually rather cheap to save life on our planet. Given the

wide range of services that we rely on from nature, it is certainly a lot less expensive than failing to save it.

The finely woven tapestry of life on Earth is, however, now more gravely at risk than it was, until recently, widely understood. Certainly habitat loss and the other pressures that have brought much of the planet's biodiversity to the brink of disappearance or to extinction have had a massive impact.

There is, however, a bigger threat still. The biosphere does not only interact with itself. It fundamentally also relies on the functioning of the atmosphere. And the atmosphere is changing, and changing fast.

The huge number of interactions between a huge number of species has taken millions of years to evolve. It should be treated with the respect it deserves.

2.

w

arming

world

fragile atmosphere

More fundamental for planetary well-being even than the Earth's biodiversity is its fragile atmosphere.

It is all too easy to take for granted the thin and delicate envelope of gases within which we live and breathe; the sky seems far too vast to be at risk from any damaging influences of humankind. The atmosphere is, however, vulnerable and is now at grave risk of very serious and rapid change—change that, if left unchecked, will have profound implications for life on Earth.

Viewed from above, the bluish layer of gases that cling to the surface of the Earth appears far more delicate than when viewed from Earth. Images beamed back from satellites reveal how, with increasing distance from the surface of the Earth, the air rapidly transforms from a lower layer—a dense blanket of gas—into an upper layer of diffuse blue that quickly gradates into the emptiness of space.

Most of the weather that so fundamentally shapes all conditions for life—from deserts to rain forests—is generated at the bottom of the dark-blue layer, in a column of gases that rises only about six miles above the surface. Even at the top of the world's highest mountains, only about five and a half miles above sea level, the air is thin enough to make breathing very difficult. The diameter of our planet is 8,000 miles, thus making the atmosphere as thin as a coat of paint on a football.

The changes that have taken place in the last few decades and which are continuing, are set to accelerate into the future. Such changes are without recent precedent, certainly in terms of the speed at which they are taking place and, potentially, also in terms of their magnitude.

In 2005, the Caribbean had its worst hurricane season ever—culminating in Hurricane Katrina. Katrina making landfall over New Orleans.

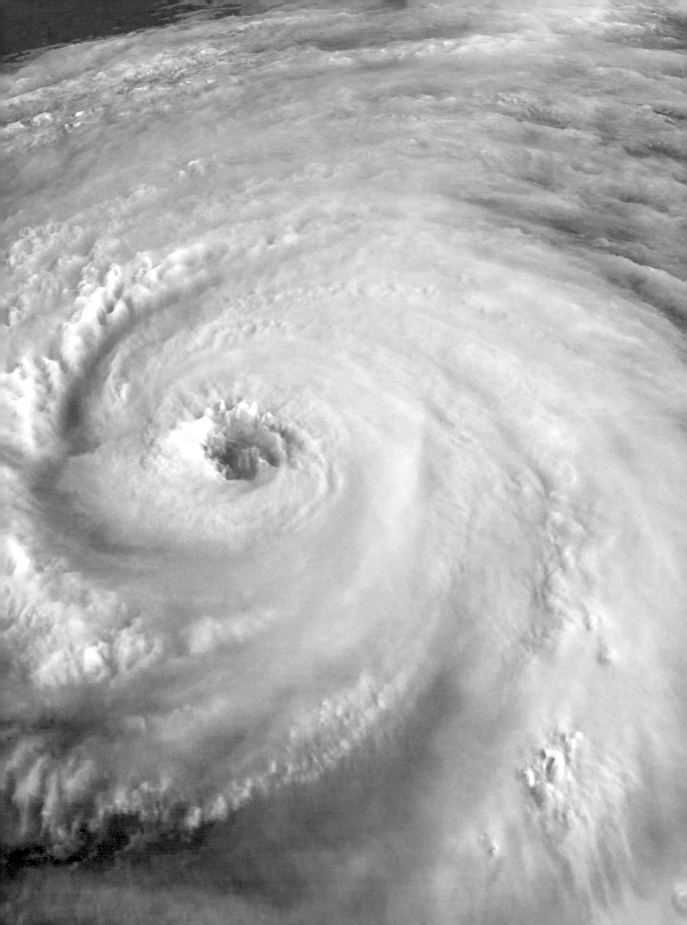

greenhouse gases

Humankind has made several impacts on the atmosphere recently, including: the release of the pollution that causes acid rain; and the increased release of manmade compounds that led to a drastic thinning in the planet's high-atmosphere protective shield of ozone. While some action has been taken to control both of these serious ecological challenges, there is now an even more major threat. It relates to the rapid increase in the concentration of so-called greenhouse gases.

The atmosphere of our planet is made up mainly of inert gaseous nitrogen: it comprises nearly four-fifths of the air we breathe. Oxygen that is essential for animal life makes up much of the final fifth while the tiny fraction that remains is made up of various gases that are naturally found in the atmosphere in low concentrations, including the inert gas argon, as well as methane, water vapor, and carbon dioxide.

These last three—methane, water vapor, and carbon dioxide—are greenhouse gases, and these keep conditions warmer than would be the case in their absence. This is a process known as the greenhouse effect. Solar energy that reaches the surface of the Earth is absorbed to a greater or lesser extent and then radiates heat back into the air. Some of that heat is trapped by greenhouse gases in the atmosphere. Without some level of greenhouse gases our planet would be an average of about 15 degrees centigrade colder than at present. This would render conditions for life extremely hostile and lead to the world being encased in a permanent layer of ice. These gases are therefore essential for maintaining the conditions that exist on Earth today. The present changes to the atmosphere are, thus, not so much to do with the presence or absence of greenhouse gases, but much more to do with their rapidly altering relative concentrations.

The Earth's thermostat is set by the complex interaction of a series of delicately poised natural forces. These include the manner in which the orbit of the Earth is influenced by the gravitational pull of Jupiter and Saturn, taking our planet

Volcanoes can emit huge amounts of carbon dioxide and sulfur compounds. Mt. St Helen's eruption on the May 18, 1980, blew ash, gas, and rock 12 miles into the atmosphere.

alternately closer to and then more distant from the Sun on a more or less 100,000-year cycle. There are also periodic wobbles in the axis on which the planet spins, which can also lead to temperature change. The impacts that result from these cycles appear to include at least some of the ice ages over the last few million years.

Overlaid on these cycles are other factors, including volcanic activity, which causes changes in levels of sulfur particles (that tend to block sunlight and cool the planet), as well as carbon dioxide (which traps heat thereby warming it up). Changing concentrations of greenhouse gases are also, in part, determined by the state of the biosphere. The main factor affecting the composition of our planet's atmosphere now, however, is human activity.

the carbon age

During the latter part of the eighteenth century momentous transformations were taking place. A combination of major breakthroughs in engineering, rapid mechanization in farming and manufacturing, new patterns of investment, workforce movement from rural areas to new factories in the cities, and the social and economic forces that were consequences of all this, are usually referred to as the Industrial Revolution. It started in England, and spread to Europe and then to North America. More than 200 years later the wave of industrialization is still spreading, more recently across China, India, and other rapidly expanding economies.

One aspect that is common to this revolution, whenever or wherever it has occurred, is a dramatic increase in the demand for energy. This shift from human and animal labor, or power provided by wood, water, and wind, to rapid industrialization has always led to an increase in the use of fossil fuels. These rich energy deposits, which accumulated in the Earth's crust tens or even hundreds of millions of years ago, now power the modern world, and have become increasingly important as industrial economies have emerged worldwide.

The original Industrial Revolution, in common with its modern parallels, was initially powered by coal. This ancient fuel source mainly originates from the vast forests that crowded the warm humid lands of the Carboniferous period (354–290 million years ago). Some coal measures are more recent, including those from the Permian era that followed the Carboniferous. There are also still more recent coal deposits, sometimes called brown coal.

The remains of ancient forests accumulated in swamps, were flooded and, under marine conditions, rocks—typically limestone, sandstone, and shale, depending on the particular circumstances, location, or time—were laid down on top of them. The compaction of the plants under the crushing pressure of rock over millions of years led to the formation of what we now know as coal. In some parts of the world these carboniferous deposits are about two and a half miles thick, the result of more than 60 million years of

The coal economy depends on burning carbon that was deposited mainly in the Carboniferous period—300 million years ago. The coal is pertrified plants that captured the carbon from the atmosphere and locked it away in rocks, until re-released when burnt. Coal fossil from Pittsburg, USA.

erosion and deposition, but the layers of coal within a vast spread of rock strata vary enormously in thickness. Some coal seams form a mere thin black line in the rock; others are massive, in places up to 200 feet thick.

Coal deposits occur throughout northern Europe, widely in Asia, North and South America, and Australia, thus providing the basis for industrial development in most regions of the world. Wherever these coal deposits are, they retain carbon that was once cycling in the ancient atmosphere and biosphere. Photosynthesis, taking place all those millions of years ago, also took billions of tons of carbon dioxide and converted it into plant biomass. Until recently, therefore, all that carbon was, in an ecological sense, out of circulation, locked tightly within geological deposits.

Another fossil source of carbon that is fast coming back into atmospheric circulation is derived from oil. Oil that naturally seeped through rock fissures to the surface was used by the Sumerians, Assyrians, and Babylonians in the ancient Middle East as many as 5,000 years ago, for, among other things, road construction and to waterproof ships. During the twentieth century, although industrial development was initially driven by coal, liquid fossil energy also came on the scene in a vast way. The modern age of oil began with a few small fields producing modest levels of energy supply. Today, the world relies on vast quantities of oil to power transportation networks and to fuel industrial farming.

As with coal, oil deposits are mainly found sandwiched in layers of sedimentary rocks, mostly in sandstone and limestone. The oil that we rely on so fundamentally today was derived from plants and animals that lived in the oceans between about 160 and 10 million years ago. When they died, they sank to the sea floor and were buried under mineral deposits. This prevented them from decaying immediately, and their remains were incorporated into these deposits over time, the whole thing becoming the source rock for crude oil. This source rock was in turn buried under successively thicker layers of sediment, which

coal-fired power stations are some of the biggest producers of green-house gases. This one is being part fired with biomass so as to reduce its overall carbon emissions.

created immense heat and pressure, and transformed the remains of the creatures into today's most vital energy source. As was the case with coal, during this process, a vast quantity of carbon that was previously cycling in the environment was trapped.

More recently still, a third fossil fuel has come into large-scale use—the natural gas that is frequently found with oil deposits. There are other fossil energy deposits, such as lignite, bitumen, and tar sands, but these are at present less important than the main three power sources which are now the mainstay of economic development worldwide.

Because they are comprised of carbon and hydrogen (coal contains the most carbon, gas the least), fossil energy sources are referred to as hydrocarbons. When they are combusted, for example, in power stations, jet engines, or gas boilers, they not only release energy but also water and carbon dioxide.

Our use of these fossil fuels today is very widespread. We depend on oil for 90 percent of our transportation, to produce food, and to manufacture pharmaceuticals and other chemicals. A vast quantity of coal is combusted to generate electricity, while more is used directly in industry, for example to smelt iron ore. And there are global consequences.

Fossil energy is the principal source of carbon dioxide emissions; in 1990 these reached 20 billion tons annually. By 2004 it had reached more than 27 billion tons. Energy demand is still soaring upward, however, and world carbon-dioxide emissions from fossil fuels are at present expected to increase to about 33 billion tons per year by 2020. Compared with 1990 levels, this is a 65 percent increase. As we shall see, this upsurge in carbon dioxide emissions is leading to striking changes in the atmosphere, which are of vital importance to life on Earth. Most importantly, it is heading toward a significant enhancement of the natural greenhouse effect.

These are not, however, the only sources of carbon dioxide to enter the atmosphere as a result of human activities. Changes taking place in the biosphere have made, and continue to make, a huge contribution as well.

at sea

It is not only on land where there is a dynamic interaction with the atmosphere. Over the last 200 years it is estimated that about half of the carbon dioxide released from fossil energy use and cement production (another major carbon-dioxide source) has been taken up by the oceans. Some dissolves into seawater. A lesser proportion is soaked up by biological activity, for example, as plant plankton make food by photosynthesis, and as animals use carbon to make shells. As these plants and animals grow and die, they can take carbon dioxide into the deep oceans.

In addition to carbon dioxide, the oceans also absorb most of the heat that has been trapped in the atmosphere because of human activities. Temperature records going back to 1961 show how the average temperatures of the oceans are increasing to a depth of 10,000 feet. And the warming that has actually taken place closely matches predictive computer models that set out what should be expected, making it clear how they will continue to warm in the future.

Study of this complex set of sources and sinks for carbon dioxide (and indeed heat) determine whether carbon dioxide levels in the air are going up or down, and thus whether the Earth is heating up. Disentangling this web of carbon and heat connections is a difficult task. Whatever the detailed explanation in terms of the exact role of forests, soils, and the oceans is, however, there is one thing we do know for sure: carbon dioxide concentrations in the atmosphere are going up, and going up fast.

the changing atmosphere

In 1957 scientists began systematically to measure carbon dioxide concentrations in the Earth's atmosphere. So as to avoid any bias that might be caused by nearby industry or urban development, they selected one of the most remote places in the planet from which to conduct their investigations: Mauna Loa in the Hawaiian Islands in the Pacific Ocean.

These measurements were begun because it was known that carbon dioxide possessed heat-trapping properties. Indeed, it has been known since the early nineteenth century that carbon dioxide is a greenhouse gas, and the notion of global warming goes back almost as long.

Scientists studying atmospheric chemistry during Victorian times variously proposed that carbon dioxide concentrations might be linked to the ice ages, and also potentially to the Earth warming up, should levels rise significantly. A Swedish chemist called Svante Arrhenius even attempted to put a figure on the kinds of temperature increases that might be expected. He postulated that a doubling in carbon dioxide concentrations might lead to about a 5–6 degree centigrade average increase. Considering the time in which he was conducting his research (the late nineteenth century), and the fact that there were no computer models, let alone advanced atmospheric theories, he was, as we shall see, remarkably close to the best-guess estimate arrived at by modern-day climate scientists.

Since the late 1950s and the first systematic collection of carbon-dioxide data, a clear pattern has emerged as to the overall trend of carbon-dioxide levels in the atmosphere arising from fossil-fuel combustion, deforestation, and other land-use changes. The line plotted from the measurements gathered at Mauna Loa shows an annual rise and fall in carbon dioxide concentrations.

Each year the levels fall back as carbon dioxide is "inhaled" by the pulse of plant growth that sweeps over the northern hemisphere in spring and summer. When the Earth tilts back again and fall and winter arrive, carbon dioxide is "exhaled" as dead leaves and other plant remains decompose. These effects

This close-up view of an ice core showes the bubbles of air trapped when the ice was deposited 800,000 years ago. Analysis of the air has shown the changes in carbon dioxide concentration over the past million years.

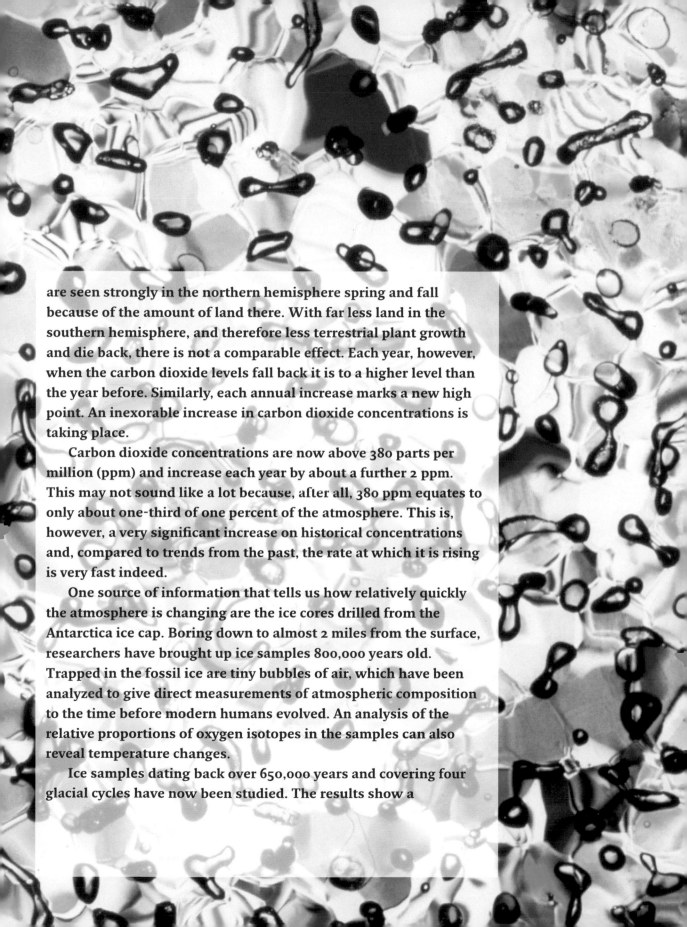

are seen strongly in the northern hemisphere spring and fall because of the amount of land there. With far less land in the southern hemisphere, and therefore less terrestrial plant growth and die back, there is not a comparable effect. Each year, however, when the carbon dioxide levels fall back it is to a higher level than the year before. Similarly, each annual increase marks a new high point. An inexorable increase in carbon dioxide concentrations is taking place.

Carbon dioxide concentrations are now above 380 parts per million (ppm) and increase each year by about a further 2 ppm. This may not sound like a lot because, after all, 380 ppm equates to only about one-third of one percent of the atmosphere. This is, however, a very significant increase on historical concentrations and, compared to trends from the past, the rate at which it is rising is very fast indeed.

One source of information that tells us how relatively quickly the atmosphere is changing are the ice cores drilled from the Antarctica ice cap. Boring down to almost 2 miles from the surface, researchers have brought up ice samples 800,000 years old. Trapped in the fossil ice are tiny bubbles of air, which have been analyzed to give direct measurements of atmospheric composition to the time before modern humans evolved. An analysis of the relative proportions of oxygen isotopes in the samples can also reveal temperature changes.

Ice samples dating back over 650,000 years and covering four glacial cycles have now been studied. The results show a

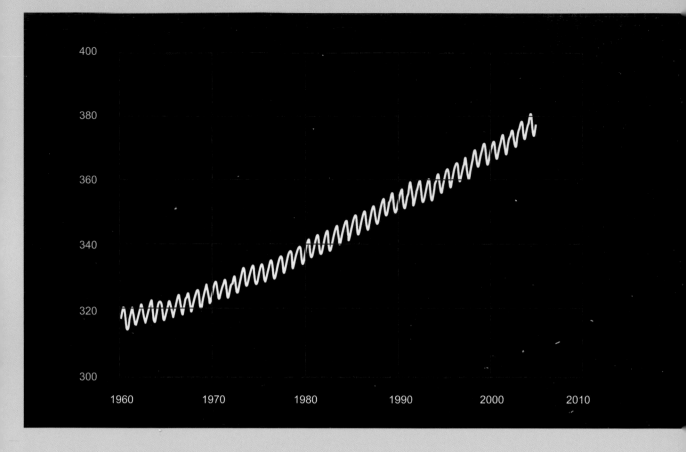

remarkably close correlation between carbon dioxide levels and
global average temperature, including the patterns of the ice ages.
When it was warmer between the glaciations, there was more
carbon dioxide; when the ice spread and it was cooler, there was
much less.

What is also remarkable from these data, however, is how
carbon dioxide concentrations over this long period never went
above 300 ppm, even in the warmer interglacial periods, and often
fell as low as 180 ppm. In the period before the industrial age
began, concentrations were at about 280 ppm. The level we have
now—above 380 and rising—is an enormous difference, and it is a
change that has occurred very rapidly.

Carbon dioxide
concentrations from
Mauna Loa, Hawaii.

other gases in the greenhouse

The warming effect caused by increased carbon dioxide concentrations is not the only source of planetary warming. It is augmented by the warming effect of other greenhouse gases. These include methane, nitrous oxide, and various so-called industrial gases, such as the chlorofluorocarbons (CFCs), which are also mainly responsible for the depletion of the ozone layer.

methane

Methane occurs at much lower levels in the air than carbon dioxide. It is, however, a far more powerful greenhouse gas with, molecule for molecule, the capacity to trap 25 times as much heat. The fact that methane levels have gone up quite dramatically is therefore an additional cause for concern alongside elevated carbon dioxide levels. Pre-industrial methane concentrations were at about 700 parts per billion (ppb) but, today, that has rocketed up by about one-and-a-half times—to around 1,700 ppb.

The rotting of organic matter in wetlands, bogs, and termite mounds were the principal sources of methane in pre-industrial times. These have been supplemented by a range of human sources, which are now responsible for about 70 percent of the total annual emissions of this gas.

The main human-derived sources of methane appear to be different kinds of agricultural activities: rice paddies produce large quantities of methane; so do farm animals, such as cattle and sheep (as a by-product of their digestion). The clearance of forests and their replacement by grasslands, that in turn have more termite mounds than the forests, is a further source of methane. Other major contributions come from the rotting organic matter in landfill sites, sewage, and the escape of natural gas from oil- and gas-production facilities.

Methane only stays in the atmosphere for about nine years, however, in comparison to carbon dioxide, which remains for about a century.

termites are one of the major natural producers of methane.

nitrous oxide

Nitrous oxide is another powerful greenhouse gas. It is emitted into the atmosphere mainly from the activities of microbes in the soil. Nitrous-oxide levels are in large part determined by land-use changes, with emissions enhanced under warm and wet conditions, and when nitrogen fertilizers are applied in farming activities. Tropical forest clearance and logging activities influence nitrous-oxide concentrations, as does the manner in which farming is carried out. More intensive chemical farming makes a bigger impact.

Nitrous-oxide concentrations have risen from a pre-industrial level of around 270 ppb to 319 ppb in 2005. Although occurring at low concentrations, lower still than the traces of methane in the air, even a small increase in nitrous-oxide levels is important because each molecule of this gas is more than 300 times more powerful as a warming agent than a single molecule of carbon dioxide.

hydrofluorocarbons (hfcs)

The different industrial gases, used for example in refrigeration, are not naturally occurring substances and, thus, are much more under our direct control than those that can change in concentration because of more natural factors. Although measured in parts per trillion (that is, at very low concentrations) they are nonetheless extremely powerful greenhouse gases. In the case of HFCs, some thousands of times more potent than carbon dioxide.

Taking stock of all the greenhouse gases together, carbon dioxide is still the main heat-trapping agent building up in the atmosphere as a result of human actions. It is thus carbon-dioxide emissions that must be urgently controlled if we are to avoid high levels of warming on the Earth in the near future. And it seems that we need to cut back emissions of carbon dioxide by even more than was previously believed necessary. This is, in part, because other pollution that causes the atmosphere to cool down is being removed.

pollution paradox

In recent decades strenuous efforts have been made in the cleanup of the power stations and industrial facilities which were causing the pollution that leads to acid rain. This was an important breakthrough, and has helped protect biodiversity. These pollution control policies do, however, have implications for climate change.

The particles of sulfur that are responsible for the acidification of lakes and forests reflect the Sun's energy while they are floating in the atmosphere, and thus help keep the planet a little cooler than it would otherwise be. White smoke has a similar effect, reflecting energy back into space before it can be absorbed by the land and sea (black smoke has the opposite effect, absorbing energy and heating up the atmosphere).

When sulfate gases combine with water to form droplets of sulfuric acid, these can act as a catalyst for cloud formation. Extra clouds can then reflect even more sunlight, thus adding a further cooling effect. The lifetime of sulfur particles and droplets of sulfuric acid in the atmosphere is very short, however, at just a few days. So the cooling effect that is generated in this way is regional, and concentrated around the industrialized areas that are the main pollution sources.

In the developed, industrialized countries the cooling effect of sulfur was greatest in the 1970s–1980s and is now declining because of successful (and environmentally necessary) policies to cut acid pollution and thus sulfur. Conversely, the burning of more coal in newly industrializing countries is leading to an increase in this pollution source, at least temporarily. The magnitude of the cooling effect of the sulfur is not clear, though, in some areas with high levels of industrial pollution such as central Europe and China, it could be a considerable influence in keeping down temperatures.

Dust going into the atmosphere because of soil erosion and desertification can also curb local warming effects. In any event, and despite the uncertainties, it is ever more clear that these and other cooling influences are nowhere near sufficient to compensate for the warming effect being caused by greenhouse gases. But how much warming can be expected and by when?

warmer times

We know for certain that specific gases trap heat. We know the gases that do this are increasing in the atmosphere. What we do not know, and what scientists are now grappling with, is what temperature changes will come with different levels of emissions of greenhouse gases.

The most recent projections published by the Intergovernmental Panel on Climate Change (IPCC) set out the latest results from complex computer modeling studies that simulate how the climate system will respond to different levels of greenhouse-gas pollution. These projections can never be completely predictive because of the huge complexity of the climate system and because, no matter how sophisticated the modeling, no one can predict the future. One of the major uncertainties in this modeling is the role of clouds. In a warmer world there will be more evaporation and more water vapor but it is also the kinds of clouds that form as well as the number that will be important, and there is, as yet, insufficient information on what is likely to happen.

Despite some obvious uncertainty, climate change modelers take more and more confidence as the credibility of their work is proved by how earlier models have matched quite well the actual warming of the atmosphere.

Taking the information available and using the most sophisticated analyses available, including a range of new cutting-edge computer simulations, climate scientists have recently agreed that there could be very significant temperature increases caused by human activities during the course of the twenty-first century. Indeed, the most recent assessment by the global panel of leading climatologists suggests that, in a worst case scenario, average temperatures could soar to more six degrees higher than the pre-industrial average.

Some researchers believe that it was this kind of temperature change that initiated the massive end-Permian extinction of just over 250 million years ago. Strikingly, there is evidence suggesting that at that time it took thousands of years for temperatures to increase, rather than the single century it could take now.

Clouds have a key role in influencing the temperature of the Earth.

As far as can be seen, the level of warming that occurs will most significantly be determined by how much atmospheric change occurs. Looking at a scenario in which carbon dioxide concentrations rise to about 550 ppm (roughly doubling the prevailing level of pre-industrial times), it is expected that the temperature will increase by between two and four-and-a-half degrees. A "best guess" put forward by scientists in this possible future is a temperature increase of about three degrees centigrade. This is a bit lower than the estimate of Svante Arrhenius, but he was in the right ballpark.

This warming that is projected in the models won't occur evenly across the world. Large regional variations are expected. It is anticipated that some will go a lot warmer than the global average: the Arctic, for example.

This three-degree increase is, in some ways, a "business as usual" scenario, and is based on us carrying on as we are now, releasing large quantities of greenhouse gases. But, as we shall see later, this level of warming can be expected to lead to very serious changes with major implications for people and wildlife.

how much warming is too much?

Even the modest level of global warming that has already occurred because of human activities—about 0.7 degrees centigrade global average increase compared to the nineteenth century—is causing major changes to the functioning of ecosystems and, indeed, weather patterns.

Changes to microclimates and larger-scale changes to weather systems have been documented in the increasing intensity of tropical cyclones and the prevalence of flooding and droughts that are occurring broadly in line with the projections set out in climate-change models. In other words, climate change is already with us, and it is proceeding largely as expected.

The key question now is not so much whether the Earth is getting warmer, but what level of warming we must seek to avoid.

There is enormous lag in the climate system because it is so vast. Even if we were able to somehow end the emissions of greenhouse gases right away, there would still be some more warming. This is because of a time delay between the release of the pollution and the warming effect it will ultimately cause. It is in some ways like placing a kettle on a gas stove: the water does not immediately boil but gradually heats up; when the gas is turned off, the kettle only gradually stops boiling. So it goes with greenhouse gases released into the atmosphere. Indeed, there is already around a further 0.6 degrees of warming in the climate system, no matter what we do in the future.

With the Earth already committed to at least 1.3 degrees of warming, therefore, and with more being added each year, it is clear that the world is changing very fast, certainly compared to any known examples from the last ten thousand years and probably beyond.

Under this circumstance of rapid change, climate scientists have started to assess what level of global warming might be deemed "dangerous"—from the perspective of the natural world, human welfare, and the economy. The emerging consensus view, now shared by many governments and environmental

Desertification—the growth of deserts around the world—is a possible feature of the increased temperature of the Earth.

organizations, such as Friends of the Earth, is that we should take immediate steps to avoid global average temperatures increasing above two degrees centigrade compared to the pre-industrial average (around 1800 C.E.). Above this level of warming it is projected that some very serious consequences would result, including a high level of species loss, large-scale impacts on agriculture, serious economic damage, and the stimulation of potentially devastating feedbacks (see later for more on this).

So how long do we have before we commit our planet to that level of warming? The growing point of view is that the level of pollution likely to push the planet over that critical threshold of two degrees is very close. One recent analysis suggests that, for the Earth to stand about a 50:50 chance of remaining below two degrees of warming, we need to peak concentrations of greenhouse gases at no more than about 450 ppm carbon dioxide equivalent (that is, the warming effect of all the greenhouse gases calculated at a carbon-dioxide equivalent) and then reduce back down from this high level as quickly as possible. This would require an aggressive and ambitious, but achievable, program of greenhouse-gas reduction.

The concentration of carbon dioxide is now escalating rapidly as the world consumes more and more fossil energy and as deforestation continues, especially in the tropics. Carbon dioxide concentrations have already risen by about 100 ppm: the first 50 ppm increase took place in about 200 years, from the start of the Industrial Revolution to around 1973; the next 50 ppm increase took place in just 33 years, from 1973 to 2006. On the basis of plausible projections set out on the strength of recent and projected energy demand, we could climb to near 600 ppm (that is another 220 ppm or so) by about 2050, and far higher even than that by the end of the twenty-first century. That is many times faster than the historical trend, and there is little doubt that the consequence of that rate and scale of growth would be disaster.

Already the atmosphere has about a 430 ppm carbon dioxide equivalent concentration of greenhouse gases and is going up at about 2 ppm per year. The gap between now and 450 ppm is thus about a decade.

No one can say for sure whether the critical level is 450, 460, or even 500 ppm. Then again, it might be 420 or even 400 ppm (in other words already passed). Should we take the prudent route and cut the release of greenhouse gases right away, or should we allow concentrations to run higher, say up to 550 ppm, before taking action? At that level, it is estimated that there is an 80:20 chance of exceeding two degrees of average global temperature increase. The latest best guess from climate-change science is that this level of greenhouse gas concentrations would lead to three degrees of global warming: in other words, far above what is desirable.

In making choices about how to respond to climate change we must assess probabilities, statistics, and make risk assessments. For society, the issue now is how risky do we want to be. This could prove to be the most momentous question ever to have faced humankind.

life in the hothouse

While public debate rages as to the causes and consequences of climate change, the science is actually very clear about the basics: the Earth is heating up, and fast, with most of the recent warming directly ascribed to human activities. If we do not take action very soon, then it is likely that a series of mostly unpleasant consequences will follow. The choice for us now is how much warming we wish to bring about. Simply put, the more pollution that we release, the hotter it will get.

effects on biodiversity

Of great concern is the impact rapid warming will cause to biodiversity. The pressures of habitat degradation and fragmentation could be dramatically worsened as changing climatic conditions cause profound ecological transformations. This will result from a variety of factors, including hotter temperatures causing plants to grow differently, with extreme heat causing die back in many species; changing moisture availability altering the distribution of animal life and affecting predator-and-prey relationships; the warming of lakes affecting their ecology and water quality; and the timing of seasonal changes putting breeding cycles out of kilter with available food sources for offspring.

Several species have already been reported recently extinct due to changing climatic conditions: the golden toad that was once found in the Monteverde cloud forests of Costa Rica is one. This gorgeous yellow amphibian lived in high-mountain tropical forests that were bathed in cloud, thus providing a particular microclimate of temperature and moisture in which the toads thrived. Increasingly, however, the clouds formed higher up, thus changing the conditions in which they lived, and the last living animal was seen in 1989.

One animal that is under pressure, and which could be rendered extinct because of the existing modest level of warming that has already occurred, is the pika. These little members of the rabbit family live in the highland areas of the western USA and

The golden toad has already become extinct, but global changes also threaten temperate animals like the pika of North America.

Canada, and already are suffering the fragmentation of their rocky habitat as the temperate zone they occupy moves toward the tops of the mountains.

Undoubtedly, there will be surprises that will cause large-scale impact on biodiversity, and over short timescales. For example, the recent near-total breeding failure at sea bird colonies around some of the northern coasts of Britain has been linked to increased sea temperatures that have affected the distribution of the plankton, the food for the sand eels, which are, in turn, the main food source for the chicks of many maritime birds. Similar consequences for sea-bird colonies, linked to increased sea-surface temperatures, have been reported from other regions as well, such as the Great Barrier Reef off the Australian coast.

The disappearance of suitable habitat is likely to be the major factor affecting biodiversity. As habitats change, and especially as already fragmented areas alter in response to warming, then extinction levels are expected to rocket. This will, of course, be an especially strong factor for animals and plants that cannot find suitable areas of new habitat as conditions change. The trend will be for habitat types to migrate toward the poles and higher up on mountains. Some will disappear, either going off the end of the world or over the tops of peaks.

The polar bear is threatened by the decrease in the Arctic spring ice. The bears now have to swim further to reach floes where they can hunt seal.

The fate of the polar bear could be decided in this way as suitable floating ice disappears from the top of the planet. Polar bears are threatened by pollution, including the build-up of highly persistent toxic substances that affect their hormonal systems, as well as new oil development. But it is now widely believed that the main danger facing these most impressive carnivores is rapid global warming. The most immediate issue is the loss of the summer sea ice on which the animals need to sit in order to hunt ringed seals, their main prey.

There are at present some 22,000 polar bears living in the Arctic, but that number can be expected to drop dramatically in the years ahead. Less ice means the bears' hunting season is curtailed, making it impossible for many of them to find sufficient food to survive. By the time they migrate to land, their fat reserves are too low, in turn reducing their ability to breed successfully. Although polar bears are excellent swimmers, quite able to cover long distances on the open sea, drowned animals have recently been found far from land. This is a new development.

Other large polar mammals are also threatened by changes to the sea ice. These include several species of whale, including the beluga, narwhal, bowhead, and right whales. Other whale species could face threats arising from the change in ocean currents and from food shortages. Having narrowly survived the effects of centuries of excessive hunting, it seems that several species of these giants of the deep are now facing extinction because of human-induced climate change.

While the pressure on some species (which arises from the very rapid increase in temperatures in the Arctic) is quite visible to researchers, the scale of the wider threat posed to biodiversity is to some extent uncertain. As is the case with climate-change models, the future projections of species loss are, to an extent, educated guesses. There is, however, every good reason to be alarmed.

One recent study examined a range of biodiversity-rich ecosystems covering 20 percent of the Earth's land area and assessed their vulnerability. Computer simulations were put together in order to understand how the ranges of more than one thousand species of mammals, birds, amphibians, insects, and plants would change in response to climatic warming. Like the climatologists who have been working to map the future climate under different pollution-level scenarios, the biodiversity researchers looked at several different possible futures: a low-warming future; a mid-range scenario; and a possible future in which high increases in average temperatures take place.

The conclusions were dramatic. Of the species studied, the

researchers concluded that some 15–37 percent would be extinct by 2050 because of climate change. Extrapolating the results from this sample of animals and plants, and applying the findings to the natural world as a whole, the findings suggest that as many as a million species will be lost—and that's the mid-range scenario.

marine biodiversity

There are also likely to be serious impacts for marine biodiversity . Factors causing the degradation of coral reefs, as we saw in the last chapter, include pollution, sedimentation, and the effects of direct exploitation, including fishing and harvesting of coral. To this list of pressures must now be added climate change.

Warming sea temperatures have already caused damage to coral reefs and there is clearly the potential for much more widespread and serious harm in the near future. Indeed, it is expected that much of the world's coral reefs will be lost, even with quite modest levels of global warming—modest, that is, compared with what might plausibly occur during the twenty-first century.

The main visible effect of warmer seas on coral reefs is coral bleaching. This process occurs when corals are stressed, which can come about with a decline in available light or increase in temperature, for example. This then causes the symbiotic relationship between the coral polyps and the tiny algae that live with them to break down. In the light of predictions that temperatures will continue to increase as a result of global climate change, there is growing concern about the future of tropical coral reefs.

Bleaching is a kind of slow reef death when the once multicolored corals that form the structure of these amazing ecosystems become a translucent and ghostly white. Several major bleaching events (linked to increased sea temperatures) have occurred in recent years. In 1998 and 2002 major bleaching events occurred in the waters of the Great Barrier Reef Marine Park, raising concerns about the long-term conservation of the reef. An aerial survey of more than 640 individual reef areas during the

Corals are extremely sensitive to heat and die if their environment becomes too hot. This so-called coral bleaching has become a worldwide phenomenon in the past decade.

bleaching event of 2002 found that nearly 55 percent of those areas surveyed showed some degree of bleaching as a result of heat stress.

The Barrier Reef Marine Park is the largest protected area on Earth, and crucial for the survival of one of the most incredible ecosystems of all. And yet, the legal protection enjoyed by the reef is of no use when it comes to rapid global warming. As we shall see shortly, the second-biggest protected area—a vast area of the Colombian Amazon—could be similarly redundant when it comes to this new threat.

Maximum summer sea temperatures just two or three degrees above normal can kill corals, whether they are inside a protected marine park or not. The upper temperature limit for corals varies between species and places. Species that usually live in cooler conditions, where summer maximum temperatures are 82°F, will bleach at lower temperatures than corals that usually live in hotter parts of the reef where summer temperatures reach 88°F. In addition to bleaching, there is also now evidence to suggest that an increase in coral disease can be attributed to rising sea temperatures.

Coral bleaching has indirect implications for other marine groups too. The loss of reef structure because of a decline in coral cover has been shown to adversely affect the abundance of reef fish, and is likely to have a similar impact on other reef-dependent organisms. Even if we manage to limit human-induced warming to below the two-degrees centigrade average increase, we can still expect to lose much of the coral reefs and the multitude of species they support.

The short-term (which means decades) prognosis is serious for coral reefs, with major reductions in their extent and biodiversity almost certain. Some corals may adapt, at least to small temperature changes, while others might move to cooler waters, but the overall trend in the short term is likely to be one of contraction and loss of diversity.

The longer-term (meaning centuries to millennia) prospects for these ecosystems is more encouraging, however. The fossil history demonstrates how coral reefs have remarkable resilience against severe disruption and will probably be able to adapt through natural selection as climate changes either stabilize or reverse; this will take time, however, and potentially quite a lot of it. The result will also depend in part on how the chemistry of the ocean changes in response to increased carbon dioxide in the atmosphere. As we shall see later, this will be a major factor determining ocean health irrespective of increasing temperatures.

Sunflowers damaged by a forest fire following the severe drought in Portugal in 2003.

our own future in the greenhouse

It is impossible to separate the impacts on biodiversity from the effects that will be felt by people. In the warming world it will not only be nature that has to adapt. Humankind must also prepare for changes. Even if we managed to drastically reduce the pollution that is causing temperatures to increase, there will still be other impacts that we will need to cope with.

In terms of global average temperature, some eleven of the twelve warmest years ever recorded were in the most recent twelve years. The record for the warmest year ever is now frequently broken.

heat

The gradual warming that is underway is not only causing conditions to change over time, but is also punctuated by extreme events that are, at least in part, attributable to climate change. The extreme hot summer temperatures that occurred across central Europe during 2003 were, in a statistical sense, unlikely to have occurred because of purely natural variation and were consistent with the type of summer temperatures that are expected to accompany climate change.

This hot spell caused some 35,000 premature deaths—from heat stroke and heart attacks among other things—and led to a severe economic impact on agriculture as crops failed from heat and drought. Nobody can say for sure whether this hot spell was owing to global warming caused by human activities, but it was the kind of event that computer modeling says is more likely to occur as global warming increases. By the middle of the twenty-first century, if greenhouse gases continue to accumulate, such a summer could be normal (rather than the exception). By 2100, a 2003-type summer is even expected to be cooler than the average.

Extreme heat is one expected outcome of increased average temperatures, so are more severe droughts and more intense rainstorms. As is the case with many of the record-breaking weather events of recent years, it is often not possible to attribute any single one of them to human-induced global warming (although the European heat wave of 2003 came closer than most). It is, however, possible to link the tendency for more extreme and more intense events with increasing temperatures: take the case of hurricanes.

storms

Severe tropical cyclones (called hurricanes in the Atlantic and typhoons in the Pacific) are not a new phenomenon, and are a natural weather event. In recent times, however, they have become more intense. Researchers have discovered that the storms occurring in recent decades have on average become stronger. This correlates with increased sea surface temperatures, in turn linked to the period of human-induced warming that is now underway.

If the world needed a reminder of the potential destructive force of tropical storms, then it got one in 2005 when Hurricane Katrina smashed into New Orleans. A tidal surge breached the defenses that protected built-up areas from flooding and swept across much of the city, causing billions of dollars' worth of damage, killing almost 2,000 people, and causing human misery on a scale rarely seen in developed countries. Large areas of the city were flooded to more than ten feet deep. Many who stayed in their houses drowned. Society instantly broke down and there was looting and robbery. The authorities were unable to restore order following the hurricane and it took weeks to take control of the situation. Even almost two years later, thousands of people were still unable to return home. And all this was in the world's richest country. Far worse outcomes even than this will accompany more intense weather events that hit poorer countries.

Northern-hemisphere tropical storms are also getting stronger, but more surprising, in March 2004, was the appearance of a southern-hemisphere hurricane in the Atlantic. Because of cooler conditions in the Atlantic, it had been believed that such storms could not form there south of the Equator. But one did: it was called Catarina and it plowed into the southeastern coast of Brazil. Climate-change modelers had anticipated that such a storm might one day occur, and indeed one model predicted very closely where the storm would form and where it would go. This was another example of the recent climate-change science accurately anticipating actual events before they occur and, therefore, supporting future projections.

This traffic jam was caused by the evacuation of Alabama on September 14, 2004, the year before Hurricane Katrina destroyed New Orleans.

rain

More intense rainstorms are another expected consequence of climatic warming and are another of the changes now being picked up in weather trends in different parts of the world. Warmer air means more evaporation and more moisture in the atmosphere. This can mean more clouds and thus more rain.

In 2005 one of the most intense rainfall events ever recorded hit the Indian city of Mumbai. In just one day, 37 inches of rain fell causing flash floods that killed about 1,000 people. The deluge triggered landslides and caused major property damage.

In temperate regions, too, there is strong evidence for a trend toward more intense rainfall events. In the UK, for example, there are data to show how intense rainstorms have doubled in parts of the country, and that rainfall is becoming more concentrated in a rainy season spanning fall and winter, whereas it had previously been more spread across the year. Very intense downpours that had been usually recorded about every 25 years are now reported more in the region of every six years. One event that hit the media headlines, and which is the type of occurrence that can be expected more often in the future, was the flash flood that hit Boscastle in Cornwall in the west of England in the summer of 2004. The small

Heavy monsoon rains hit Mumbai and 37 inches of rain fell in a 24-hour period (red indicates torrential rainfall). Coupled with high tides this caused heavy flooding of the city.

coastal town was devastated by a flash flood following a storm in which eight inches of rain fell in just a couple of hours.

drought and disease

Paradoxically, perhaps, climate change is also linked to more drought, including in places that are experiencing more intense rainfall as well. The implications of this situation for agriculture can be grave, as was seen in Europe in 2003, and over large parts of East Africa between 2004 and 2006. An absence of rainfall can lead to serious economic damage and, in extremes, to food shortages.

Some major public health threats are also linked to global warming. Indeed, one estimate is that around 150,000 people are already dying each year because of the warming that has already occurred, in large part because of the spread of disease. For example, malaria is spreading into areas where it was previously absent, extending higher in mountain areas, and further north and south in response to warmer temperatures.

Burchell's zebra (Equus burchelli), eating supplementary food during a severe drought. In southeastern Africa humans and the wildlife need feeding to survive.

Dead trees after a drought in Australia.

economy

Businesses and economic development will also see increasing challenges. One sector that is directly affected by climate change, and especially by extreme events, is the insurance industry. Their commercial activities rely on assessing the risk of damage to people and property and offering cover against that risk based on a careful assessment of its likelihood of occurring. In the global greenhouse this business strategy is becoming ever more difficult, not least because what has happened in the past is not necessarily an accurate guide to what might happen in the future.

The huge increase in the payouts insurance companies have made in recent years amply demonstrates the point. Even following the modest levels of recent warming, insured losses are doubling decade upon decade as extreme events take their toll on people and property. From hailstorms in southeastern Australia and floods in southern England to record numbers of tornadoes in the USA, the sums being paid out by insurers due to increasingly extreme weather are astronomical. All records were broken in 2004: that year alone saw about 75 billion dollars in insured losses. The implications for the wider economy are serious.

Insurance is a centrally important part of economic development, having a close bearing on risk assessments and, thus, investment patterns. There are also serious social equity issues linked to the availability and cost of insurance. For example, in many low-lying and coastal areas that will be increasingly susceptible to flooding, it is likely that insurance cover will become more difficult, or even impossible, to secure. This will cause especially serious consequences for poorer communities and householders.

Even more basic needs are at increasing risk of being disrupted in the global greenhouse, however.

the future for food and water

Food and water are our most essential requirements, and societies everywhere depend on nature to provide them.

effects of drought

The 2006 drought that affected much of Australia demonstrates how even one of the richest countries (in per capita terms) is vulnerable to reduced rainfall. Australia is affected by drought naturally, but some of the climate-change models project that droughts could become more prolonged and intense. One of Australia's premier scientific bodies has suggested that rainfall in parts of eastern Australia could drop by 40 percent by the year 2070 and could be accompanied by a dramatic increase in average temperatures. This combination would massively increase the risk of bush fires, which are already a threat in Australia.

The severe drought of late 2006 began with the onset of a dry period that some believe is the most intense for the last 1,000 years. The Murray-Darling river system provides more than three-quarters of the water that is consumed in the country. So, with these rivers running at more than 50 percent below their previous minimum flow, serious challenges emerged. This was not only for consumers who faced shortages in the cities, but also for farming: wheat yields for 2006 were dramatically down on previous years.

While in many industrialized countries an interruption in food production can be dealt with through the purchase and import of extra supplies, for many rural communities in developing countries, this is not so easy. As a continent, Africa lives on the front line of global warming. Nearly three-quarters of the workforce rely on mostly rain-fed agriculture for their livelihoods, and climate change is already disrupting vital rains, bringing more droughts and floods, and changing entire landscapes. It is now estimated that between 75 and 250 million Africans will suffer from increased water stress by 2020. This will be the case in the particularly semi-arid regions of the continent. The impact on rain-dependent farming is likely to be serious—up to a 50 percent

In the 35 years since 1970, Lake Chad has dried up considerably, shrinking to around one-twentieth of its, size, and the desert sand dunes have begun to encroach on its northern shore.

reduction in farm output by 2020. The recent changes taking place at Lake Chad are perhaps a window on the future.

During the 1960s Lake Chad was the fourth-largest lake in Africa, with an area of 10,040 square miles following periods of strong rains, driving speculation that the lake would take on the huge proportions witnessed by early explorers during Victorian times. The opposite occurred, however. During the 1970s there was a drop in rainfall across sub-Saharan Africa and the lake began to shrink, with dire consequences for the people who live there.

Much of the farming that takes place around the lake depends on irrigation, but, as the lake level fell, irrigation became more and more problematic to sustain. In 1973 the lake fell to such low levels that it split into two parts with dry land in between and, by 2000, it only had an area of 965 square miles. This has prevented much of the rice production which formerly sustained local populations. Whether or not the contraction of Lake Chad is linked to human-induced global warming, the reduction of available water there demonstrates the potentially dramatic impact of reduced food production on human communities.

Of course, climate change will not lead to a drop in food production everywhere. In places production will increase, at least to begin with, as the growing season lengthens and farmland becomes more productive in some cooler temperate regions. This is indeed what some climate models expect to occur.

In many of the developing tropical regions it is, by contrast, anticipated that changed conditions will reduce food production. The effect on prices and the need for poorer countries to import more of their food can be expected to have serious impacts on poverty alleviation and development.

Climate change could lead to many more people being at risk of hunger. One model suggests that a 50 percent increase is plausible during the twenty-first century, because of increased frequency and severity of droughts (an especially serious threat in semi-arid areas) and decreases in crop yields. There is also a real threat posed by the flooding of currently productive farmland caused by the rise of sea levels.

sea levels and glaciers

Modest increases in sea level have already been recorded, in part because of the melting of glaciers. Right across the world the retreat of mountain glaciers has been dramatic: from Scandinavia to Central Europe, from Africa to the Himalayas and from Australia to South America, glaciers are moving up the mountains.

This is not only an issue for the wildlife of glacial landscapes, which is losing its home, it is also leading to the loss of irreplaceable archives. Mountain glaciers, like other long-accumulating bodies of ice, hold within them precious clues about the world in the past. Some are of huge regional importance, for example the glaciers of Mount Kilimanjaro in East Africa. These unique ice formations are disappearing so fast that they are expected to have vanished completely by 2015; shortly thereafter it is expected that all of Africa's glaciers will have melted.

The Rwenzori Mountains of East Africa, also known as the Mountains of the Moon, straddle the border of Uganda and the Democratic Republic of Congo. Not only do they hold a large proportion of Africa's very few glaciers, but these mountains contain one of just four remaining tropical ice fields outside the Andes. Rising temperatures over the last forty years or so have led to a rapid retreat of the ice, and the relentless march continues.

When surveyed over a century ago the glacial cover over this mountain range was estimated to be 2.5 square miles; this receded fast with the area covered by glaciers cut in half between 1987 and 2003. Now, less than .4 square miles of glacier ice remains. It will probably all be gone by 2025. Within a short time, it seems inevitable that the only glaciers remaining on the African continent will be in memory, folklore, and culture. Ernest Hemingway's 1936 *The Snows of Kilimanjaro* will perhaps seem like an odd title for a story about the sun-baked wastes of East Africa.

It is not only for scientific, cultural, or esthetic reasons that we should be very concerned about the rapid loss of glaciers, however. The glaciers on the Mountains of the Moon are ultimately the source of the Nile River. And these ice fields are not alone in being the origin of a great river; many other glacial areas spawn major watercourses that are of direct and immediate importance not only to a vast array of biodiversity, but also to the lives of billions of people.

The glaciers that clothe the vast Himalayan plateau are of especial importance in this respect—some of the world's most

economically important rivers rise in the mountains' glacial fields: the Indus, Ganges, Brahmaputra, Mekong, Yangtze, and Yellow rivers among them. Water from these historic waterways supplies the needs of almost half the entire human population of the world, including those in its two most populous countries: China and India.

In its arid upper reaches, it is glacial meltwater that largely sustains the Indus. Replenished by snow each winter and thus providing river flow in the summer, the glaciers are vital for a huge segment of humankind. At present, 3,500 glaciers sit within the catchment of the Indus, and their progressive melting because of climate change is set to cause serious economic and social stress in the future.

Widespread melting of Himalayan glaciers is already well under way. The glacier from which Edmund Hillary and Tenzing Norgay set out on their historic march to conquer Everest for the first time in 1953 has already retreated three miles since then. And it is not alone.

The warming that has already taken place has led to the retreat

of 67 percent of the thousands of glaciers in the Himalayan region. Continued glacier melt will increase summer river flows for a few decades and will lead to an increased frequency of floods. In time, however, there will be a severe reduction in the flow of these important rivers, such as the Ganges and Indus, as the glaciers disappear. This will lead to major impacts on the vast human population dependent on these rivers, to agriculture, and to industry.

In common with the great benefits we derive from biodiversity, the services we have gained from a stable and predictable climate have been vast. All of the great human civilizations have emerged in the ten millennia following the end of the last ice age. We have enjoyed relatively stable conditions, which have enabled us to expand our numbers on a massive scale.

We have had a few glimpses of a possible future, however, seen in the costly and socially disruptive impacts that are linked with less than one degree of warming we have already caused. Future warming will undoubtedly be more that this, though. How much more is to an extent still up to us, at least in large part in our control. At some point, though, we may lose much of that control— or all of it.

Dead leopard with two nurses in the extensive snows of Kilimanjaro in 1926. This frozen, mummified leopard carcass was discovered near the summit of Mount Kilimanjaro, Tanzania, the highest mountain in Africa. It was immortalized in the 1938 short story *The Snows of Kilimanjaro*, by Ernest Hemingway.

Retreat of the Helheim glacier in Greenland between 2001 and 2005, satellite image. These three views of the glacier show it on May 12, 2001 (top), July 7, 2003, and June 19, 2005 (bottom). The glacier itself is the gray region at the left of each image. To its right is a narrow fjord filled with icebergs that have broken (calved) from the glacier's front. The glacier has retreated around 2.5 miles in the four years between the first and last images, having been in a relatively constant position until 2001.

feedback

Increasing carbon dioxide levels caused by emissions from fossil fuels and deforestation are obviously a cause for serious concern. On top of this are the additional emissions that could be unleashed by natural systems as the world warms up. In this and other respects we should not expect the accumulation of greenhouse gases to lead to a smooth and predictable increase in temperatures. There could be shocks and surprises as the world warms.

The potential for more releases of greenhouse gases as a result of the warming put in train by human activities is called "positive" or "emissions" feedback. The warmer it gets, the less control we will probably have over the final total level of warming. There are several potential feedback effects, and some of them are absolutely vast.

One is the potential die back of the Amazon rain forests. As if the depredations of loggers, ranchers, soy farming, and mining weren't serious enough in causing the degradation, fragmentation, and progressive clearance of this incredible planetary asset, it now seems that there is a far larger threat to the Earth's largest rain forest: a lack of rain.

Modeling studies carried out by climatologists show that there is a risk of reduced rainfall over the Amazon basin caused by an increase in sea-surface temperatures in the tropical Atlantic Ocean. The prolonged drought that affected much of the rain forest during 2005 was perhaps a portent of a possible future.

Drought that transforms serious once-only events into more frequent dry periods and then moves toward such conditions being normal would have very profound effects. Not only would such a shift directly affect the composition of the forest and change the animal life within it, it would also seriously increase the risk of fire and the potential for the forest to rapidly transform from dense moist jungle into savannah and even grassland. The fact there used to be lush moist forests growing across large swathes of what is now the Sahara Desert demonstrates how climatic shifts can

trigger truly awesome changes in the dominant kinds of vegetation.

We know of no certain threshold at which such a shift would certainly occur. However, if temperatures edge up toward three degrees average global increase then the collapse of the Amazon rain forests into a much drier ecosystem become far more likely. If this happens, a vast quantity of carbon dioxide would be released into the atmosphere; there are billions of tons of carbon locked into the living fabric of the Amazon rain forests and the release of

A carving in sandstone from Saharan Algeria. It dates from around 9,560 B.C.E. to 6,000 B.C.E. when the climate of the Sahara was much wetter than today and supported savannah vegetation and animals, such as this elephant.

even a fraction of that would add considerable further impetus to the process of global warming.

It's not only in the tropics that these emissions feedbacks could be stimulated on a large scale. There is grave cause for concern in the Polar regions, partly because of the melting permafrost. Across swathes of Siberia and North America, in particular, there is a vast territory of frozen land in which bubbles of methane have been locked tight for many centuries. As the global thermostat has been edged upward, the line of melting has traveled progressively farther north.

Aerial view of patterned ground with ice wedge polygons in the arctic near Hope Bay, Canada. The frozen ground contains methane that will be released in huge quantities should the tundra melt.

Russian researchers have already detected an increase in methane emissions from the land and lakes in Siberia and expect that far more will follow if temperatures continue to rise inexorably. The sub-Arctic region of western Siberia has begun to melt especially quickly and the rapid thaw of bog lands that have been frozen for thousands of years could lead to billions of tons of methane being released, itself causing more warming. One estimate suggests that the bogs in this extensive region of Russia contain about one-quarter of the world's supply of methane held on land surfaces—70 billion tons (remember that, in 2004, world annual emissions of carbon dioxide from fossil fuels were about 27 billion tons). This part of the world is warming up very fast, recording an increase of about three degrees over the last 40 years. Alaska is a similar story.

Some, or even much, of the methane might break down before it is released and escape into the atmosphere as carbon dioxide. This would cause far less global warming that if methane is the predominant gas released, but the impact would still be potentially vast.

In a fast-warming world the soil is another increasing source of carbon dioxide. The activity of microbes in soil is governed by temperature. Their metabolism and their ability to break down organic matter, and thus to release carbon dioxide stored in that organic matter, increase exponentially with temperature. In one model, soil-carbon levels begin to decline rapidly with global warming, and add to carbon emissions far faster than increasing plant growth can compensate by taking out carbon dioxide.

Agriculture has already caused a drop in soil organic matter in many regions, and this will worsen as temperatures increase. A 25-year survey of UK soils came to the remarkable conclusion that 13 million tons of carbon is being lost each year. That, even at 1990 levels, is equivalent to about eight percent of total UK carbon emissions.

Higher temperatures also have an impact on the carbon

balance above ground in the vegetation itself. One study of the impact of the 2003 heat wave and drought across Europe estimated that the carbon dioxide produced by plants (taking into account what they absorbed) was equivalent to about one-tenth of global emissions derived from fossil-fuel combustion. The heat and dry conditions caused plants to grow more slowly, or to die, leading to a far slower uptake of carbon than would have been the case during a normal summer. Researchers concluded that, during the heat wave, plants were putting more carbon dioxide into the air than they were absorbing—and on a massive scale.

Estimates show that, during a normal year, plant growth across Europe absorbs about 125 million tons of carbon; in 2003 it is reckoned that plants released about 500 million tons. In a warming world the balance between the carbon in the plants and soils could thus be rapidly altered in ways that might hasten levels of warming put in place by human activity.

sea ice

Other feedbacks are also occurring that, although not directly contributing to more greenhouse gases, are still causing temperatures to rise. One major factor in this respect is the loss of Arctic sea ice. Over recent decades there has been a thinning and reduction in the area that is covered by ice on the Arctic Ocean. The trend is for new ice to be less widespread and thinner than previous years, and for melting in spring and summer to be more rapid and extend further.

Satellite observations reveal how, since 1978, there has been an average decrease of about 2.7 percent per decade in the extent of sea ice. Summer sea ice has reduced by more than seven percent and the record for the greatest extent of summer melting has been repeatedly broken in recent years. Compared to 100 years ago there is now, on average, about 772,000 square miles less ice. This is not

The minimum extent of Arctic sea ice in 1979 (top) and 2003 (bottom).

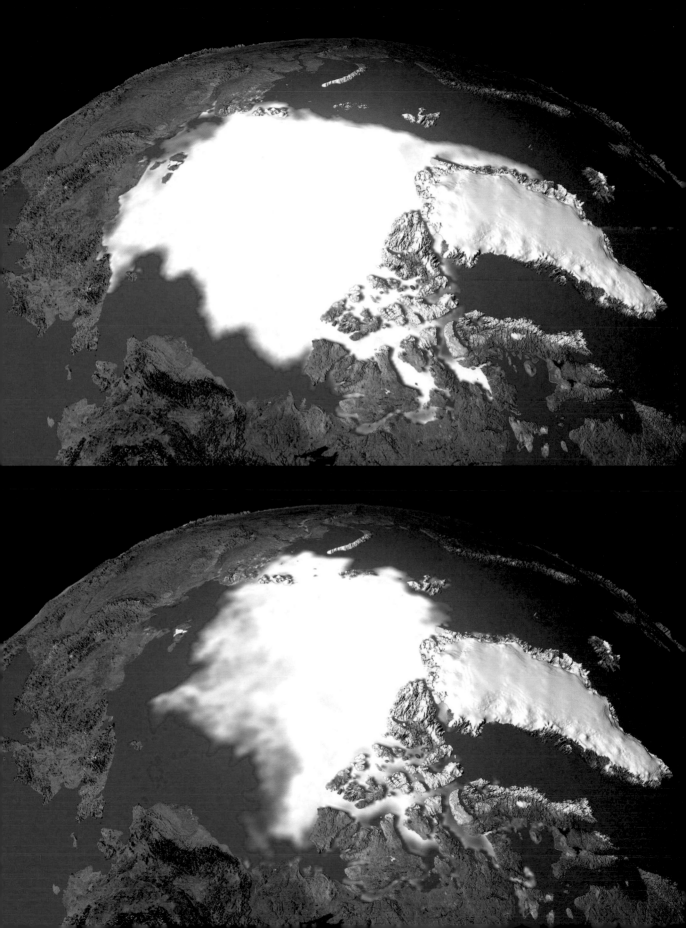

only of great importance for Arctic ecosystems; it has global ramifications.

The cap of ice on the top of the world is like a great mirror—it reflects much of the Sun's energy without warming the planet's surface. Indeed, some 90 percent of the energy hitting that huge area of white is returned to space. In contrast, the dark surfaces that are exposed when the ice melts—whether it is green ocean or dark-brown tundra—absorb much of the energy coming in. More and more of the darker surfaces are becoming exposed and the heating of the Arctic is speeding up as a result. This is one of the reasons why temperatures there have increased about twice as fast as the planetary average.

Globally there has been roughly a 10 percent reduction in ice and snow cover since the 1960s and this tendency toward a higher proportion of dark surfaces will amplify the impact of the greenhouse gases that have already accumulated and which continue to build up in the atmosphere.

At higher levels of warming it is even possible that a truly vast reservoir of methane could be released from the seabed. Methane hydrate deposits have built up from the remains of rotting sea creatures; although these can apparently remain stable for long periods, they could begin to bubble into the atmosphere if ocean temperatures increase above a certain level. The quantity of this material lying at the bottom of the oceans is massive. Should there be a large-scale release, triggered by global warming, then there could be an addition of greenhouse gases to the atmosphere that utterly dwarfs human emissions.

It is certainly in the long-term interests of humankind to avoid setting these large-scale feedbacks. It is also very much in our interests to, as far as possible, avoid triggering any significant disintegration of either of the world's two major ice caps.

awakening ice giants

The best estimate now provided by the climate change scientists as to what might be expected from the pollution now being released, and what is expected in the decades ahead, is a three-degree global average temperature increase. This does not sound high but, in terms of a global average, it is huge. If temperatures do increase to that level, very serious consequences for both nature and people will result. One major effect will be large-scale ice melt.

The most recent global assessment on the impacts to date of climate change shows a small increase in sea levels arising from ice melt. Although the sea ice that is rapidly disappearing in the Arctic has not made an impact (because it was already floating in the sea) the melting of glacial ice on land has increased the depth of the oceans by a small amount. The rise in ocean temperature has also led to a slight expansion in the volume of water, and this too has helped to elevate sea levels. So far, however, the combined effect of these factors has been relatively minor. This may not remain the case.

greenland

Locked into the icecaps on Greenland and Antarctica are huge volumes of water that, if released by melting, would have a major impact on sea levels. Already some scientists are expressing concern about the long-term stability of the Greenland icecap and are raising the alarm about the prospect that even quite modest future warming could lead to the long-term loss of much of this vast block of frozen water. The Greenland icecap holds about 700,000 cubic miles of ice—about 10 percent of the world's ice. If it all melted, it would raise the average sea level about 23 feet causing flooding of low-lying areas worldwide.

Recent survey information has confirmed that the Greenland icecap is melting much faster than previously believed. According to measurements taken by satellites, the ice mass shrank by a vast 57 cubic miles each year from April 2002 to November 2005.

The edge of the Greenland Ice Cap. The ice cap only floats on water at its edges, so like the Antarctic Ice Cap melting of this ice cap will have a major effect on the sea level.

Measurements of both the speed of glacial flow into the sea and the height of the ice determined that, between 1997 and 2003, the annual rate of melting was about 19 cubic miles. In other words, the melting has increased hugely, and over a short period.

Other researches have found that, at the edge of the ice sheet, the melting is moving even faster. Automatic monitoring stations placed at the margins of the ice show that melting is now happening ten times more quickly than earlier research had indicated. Satellite and aircraft measurements back up this finding and confirm that the margins of the Greenland ice sheet are falling in height by about three feet per year; in some places it is going even faster. The loss of Arctic sea ice is also perhaps accelerating the melting of ice on land across Greenland because of the increase in regional warming caused by the reflective white surface being progressively replaced with darker sea and land that absorb more of the Sun's energy.

It is not only on the margins of the ice sheet that there is serious cause for concern. Pools that have formed on the surface are melting through ice hundreds of yards thick to reach the rock below. This is leading to water reaching the base of the ice sheet. By lubricating the base of the ice against the rock, this could cause large chunks of glaciers to go quickly into the sea thus possibly accelerating glacial flow.

One sign to suggest that this process is already underway is the increased frequency of glacial earthquakes taking place on Greenland. Sometimes glaciers lurch forward with such force that they generate earth tremors that are detectable with seismometers right across the world. Indeed, a rapid 33-foot movement of an ice block roughly the size of Manhattan Island and as deep as the Empire State Building has caused quakes registering at a moderate five on the Richter scale.

There were about 6 to 15 quakes per year from 1993 to 2002; the number jumped to 20 in 2003, 23 in 2004, and 32 in the first 10 months of 2005. The increased quake activity matches with increased Greenland temperatures. The quakes also appear to be linked to the seasonal ice melt that lubricates glacier movement. These and other observations lead scientists to conclude that something major is taking place on Greenland and that the ice sheet there is far more vulnerable to temperature increases than was previously believed.

The question is: what will this mean for the whole Greenland ice mass, and when, if at all, will it disintegrate? Some scientists believe that, if average temperatures increase by two to three degrees in Greenland, the ice cap will begin to fall apart. Not only would this lead to dramatic increases in global sea levels, it might also affect ocean circulation by interrupting the Gulf Stream that, at present, keeps northwestern Europe far warmer than other regions at similar latitudes, such as eastern Canada and Eastern Siberia.

antarctica

Greenland is not the only potential source of major sea level rise. Antarctica could, in time, contribute as well. The southern ice continent holds nearly 90 percent of the world's ice, but, fortunately, it does not appear to be as sensitive as Greenland because temperatures there are still mostly too cold for surface melting (which accounts for about half the mass lost from the Greenland ice sheet) and because it is land surrounded by ocean, rather than ocean surrounded by land, as is the case in the northern polar region.

The potential for rapid change in conditions in Antarctica was seen in 2002, however, with the sudden break-up of the Larsen B section of the Larsen Ice Shelf. This vast ice field was attached to the Antarctic Peninsula at the most northerly part of the continent, and in a region that is warming very rapidly. Before it disintegrated, water pools similar to those increasingly observed on the surface of parts of the Greenland ice were observed forming on its surface. The break-up of 5,000 square miles was sudden and took the scientific community by surprise because the ice shelf had been quite stable in the previous 10,000 years.

Set against some of the loss of ice in Antarctica is the accumulation of extra snow, at least in some regions of the frozen continent. This is forming new ice and thus compensating for at least some of that which is being lost. Climate models have projected how warmer conditions should lead to more moisture in the atmosphere that would in turn fall as rain or snow. This does indeed seem to be what is now happening, at least in East Antarctica.

It is not so simple to see this additional snow as a direct

compensation for lost ice, however. While one satellite survey suggests that, between 1992 and 2003, the vast East Antarctic ice sheet gained some 45 billion tons of new ice, recent research suggests that what is being lost is far greater than that being accumulated—especially from the smaller West Antarctica ice fields, which are losing mass on a large scale. Glaciers there are accelerating their march toward the sea and steadily carving off more and more icebergs.

Recent surveys of the Amundsen Sea sector of West Antarctica show that glaciers are discharging about 60 cubic miles of ice per year into the ocean, which is about 60 percent more than they are accumulating in their basins from new snowfall. Moreover,

One of the largest icebergs ever recorded breaking off the Ross Ice Shelf in 2000. Over 185 miles long and 25 miles wide, it immediately ran aground on Ross Island, changing the local sea currents and winds.

measurements show that glacier-thinning rates observed during 2002 and 2003 were much greater than during the 1990s. If there is a break-up of sea ice, glaciers could flow into the sea much faster. Sea ice currently acts like a dam, slowing down the movement of glaciers into the sea: some sea ice is frozen to the seabed, making a secure block against ice running quickly off the land. If the sea ice breaks up, then the loss of land ice could accelerate.

The loss of Antarctic sea ice also has major implications for biodiversity. Krill, a kind of shrimp that occurs in great abundance in the southern oceans, is near the base of the food chains that support many larger animals, from penguins to whales and albatrosses to seals. The krill needs sea ice to complete its life cycle and a drop in krill caused by ice melt will have implications for the whole ecosystem. There is evidence pointing to a drop in krill numbers of about 80 percent in the southwestern Atlantic since 1979, which correlates with evidence that there is at least 30 days less sea ice per year at present.

While the decline in sea ice is proceeding in line with increasing temperatures, the science of ice-cap melt is still, however, uncertain. What temperature thresholds will cause rapid disintegration of land ice? What is the possible mitigating effect of more snow? What is generally accepted, however, is that extra warming will accelerate this process and lead to further sea-level rise. It seems that the warmer it gets, the greater this risk of large-scale ice-cap disintegration becomes and, the quicker it warms, the more likely it is that the risk will increase in the shorter term.

Krill are the major prey of most of the large Antarctic wildlife and are very sensitive to loss of sea ice.

using the past to estimate the future

Although we cannot predict with total certainty what will happen in the future, we can look back and see how the world was when it was previously warmer. To see what conditions were like when it was three degrees warmer, we need to look back to the Pliocene period of between five and one million years ago.

During the Pliocene epoch the world was generally warmer than at present. Warm-water plankton found in ocean deposits, and plant and animal fossils on land, give clues as to how different things were under these warmer conditions. The greatest warming seems to have been in the Arctic and cool temperate latitudes of the northern hemisphere, where temperatures were often warm enough to allow species of animals and plants to exist hundreds of miles north of the ranges of their nearest present-day relatives. Fossilized tree leaves, for example, birch leaves found in the middle of Greenland, indicate the kind of changes we can expect should temperatures increase by about three degrees.

As would be expected when forests were growing on Greenland, it seems there was far less ice volume near the poles. As a consequence, sea levels during the warmest periods in the Pliocene may have been as much as 98 feet higher than at present.

There are other ominous warnings from the more recent past too. Before the last ice age, and after the one prior to that—some 125,000 years ago—the world was about as warm as the climate models project it will be by about 2100. Sea levels then were about 20 feet higher than at present, in part because of large-scale melting of ice in Antarctica. A 20- or 23-foot increase in sea levels does, thus, seem quite plausible if temperatures increase in line with the "business as usual" scenarios put forward by the latest climate-change models. This would be disastrous.

Much of the world's most fertile agricultural land lies below this level, for example the Nile Delta and much of most productive lands of northern Europe. Many of the world's largest cities would be affected by such a sea level increase too: London, New York, Miami, Shanghai, San Francisco, and Calcutta are among them. A large proportion of entire countries would be inundated: the Netherlands and Bangladesh would be devastated, and some Pacific island nations would disappear altogether. So, at the same time as food production would be drastically diminished (by both sea level rise and drought), hundreds of millions of refugees would be on the move, looking for security, food, and shelter.

Low-lying coastal areas—coastal forests, estuaries, marshes, and dunes—are of great importance for biodiversity and will be lost if there is significant sea-level rise further impacting on many species and habitats. This kind of future used to sound like science fiction. It is now a credible scenario based on the latest science from the world's top climatologists.

The warming effect of greenhouse gases is not, however, the only challenge we need to face as a matter considerable urgency. About half of the carbon dioxide released as a result of human action over the last two hundred years or so has been absorbed by the oceans. This process continues, with a high proportion of existing emissions finishing up absorbed into seawater. This is having an effect on the acidity of the oceans. As the carbon dioxide dissolves in water, it forms carbonic acid, which turns the surface of the seas more acid.

acid oceans

There has already been a recent increase in ocean acidity, and as more and more carbon dioxide is absorbed, this is set to worsen. Emissions of carbon dioxide over the last 200 years or so have led to a strengthening in ocean acidity of about 0.1 pH units. It could intensify by as much as 0.5 units more during the twenty-first century. If this happened, oceans will be more acidic than at any time for hundreds of thousands of years and it will mark a rate of change far faster (perhaps by 100 times) than has occurred yet. Irrespective of the impacts arising from the oceans' warming, this tendency toward a more acid marine environment has profound implications.

The normal state of the oceans is slightly alkaline. This chemical environment enables shell-forming animals to thrive, so changes that lead to more acidic conditions will cause serious problems for some species. Corals are again one of the groups of animals expected to be at particular risk. Also potentially at risk are the tiny shell-forming creatures that drift in the ocean known as plankton. Although individually small, tens of billions of these animals with their carbonate shells sink to the sea floor each year, making them a significant factor in the global carbon flux. In other words, raised carbon dioxide levels in the atmosphere could make

A petrified tree from Antarctica. During the Eocene Epoch, 53–36 million years ago, Antarctica was covered in temperate forest.

the oceans acidic to the point that shell-forming animals have a diminished ability to grow shells, thereby partly limiting the ability of the oceans to absorb carbon.

There are thus serious biodiversity and climatic implications arising from ocean acidification. There are also potentially major questions for ocean productivity and the future functioning of marine fisheries. Social and economic consequences will in turn result.

cool it

If we are to be successful in conserving the Earth's biodiversity (or even a large proportion of it), it is essential that there are strenuous efforts immediately put in place to reduce the emissions of greenhouse gases. This is not only for reasons of conserving nature; if we wish to meet human needs into the future, then the same ambitious program of pollution reduction is needed.

Climate change is a global challenge and it will require a global response for it to be addressed in time. This much is now clear.

The world has known about human-induced global warming for a long time. The first awareness that some gases trap heat goes back more than 150 years. In the late 1980s scientific research was published showing how the world was indeed warming up, and shortly afterward talks were started in the United Nations in order to agree a new treaty that would seek to limit greenhouse gas build-up to manageable levels.

international treaties

In 1992 at the Rio de Janeiro Earth Summit world leaders signed up to an outline treaty that sought to orchestrate a global response. This was the United Nations Framework Convention on Climate Change (UNFCCC). Considering the vast complexity of the issue, and the fact that in the early 1990s the climate-change science was less advanced than today, these negotiations were concluded remarkably quickly. Within two years of the treaty being agreed at the Rio Summit, sufficient countries had ratified the agreement through their national parliaments for it to come into force.

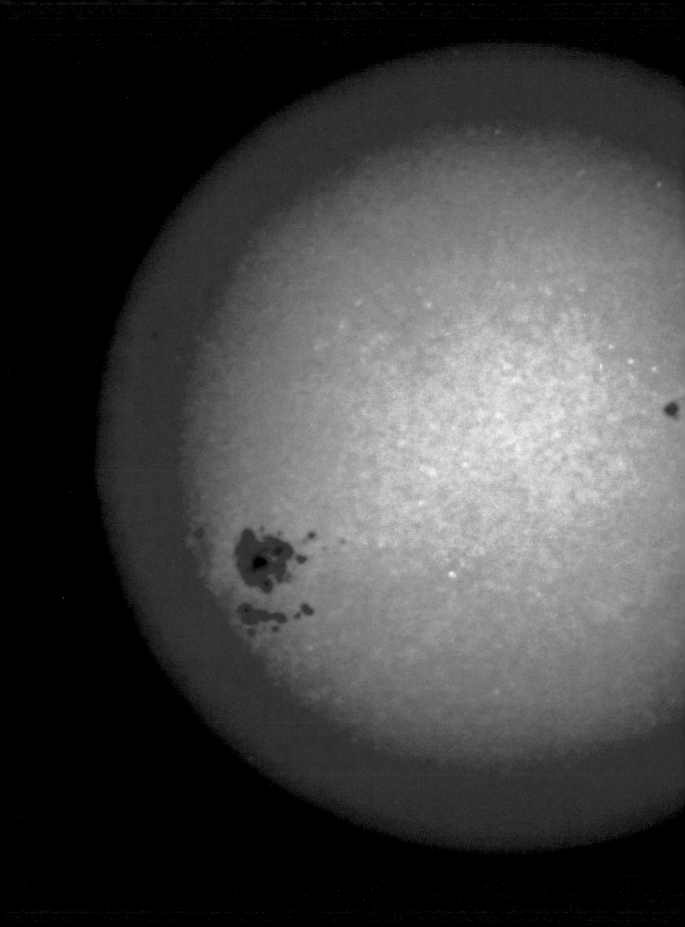

The treaty put in motion collective action by countries so as to avoid "dangerous climate change." This was interpreted in the legal text of the treaty by steps to seek the "stabilization of greenhouse gas concentrations in the atmosphere at a level that would prevent dangerous anthropogenic interference with the climate system. Such a level should be achieved within a time-frame sufficient to allow ecosystems to adapt naturally to climate change, to ensure that food production is not threatened and to enable economic development to proceed in a sustainable manner."

At first there was only a general promise by the rich, industrialized countries that are responsible for most of the greenhouse gas emissions. They undertook to stabilize how much they produced and to keep it at 1990 levels by 2000. Nearly all of them missed even this modest goal, with the exception of the UK and Germany, and both of these did it by accident rather than because of policies they had put in place to cut emissions deliberately. The former achieved it because of a shift toward gas and away from coal for electricity generation; the latter cut its pollution following the reunification of East and West Germany and the consequent decline of dirty industries that followed the collapse of communism.

Recognizing that the aim of stabilizing at 1990 pollution levels was massively insufficient, further negotiations were started in the mid-1990s that led to the agreement of the Kyoto Protocol in 1997. This new treaty was negotiated under the original United Nations agreement and did include at least modest emission-reduction targets that would be met by the major industrialized-country polluters. The EU was to cut back by eight percent compared with 1990 levels, the USA and Canada by seven percent and Japan by six. These reductions were to be achieved between 2008 and 2012. Very few countries will reach these goals, however. The USA and Canada are already way above their 1990 emissions levels and they are increasing year on year; many EU countries are in a similar position.

Sunspot activity and increased energy output from the Sun is often used to contradict the majority of scientists view that global change is caused by human activity.

Even if the Kyoto Protocol reduction targets were met, and even if emissions were kept at the new lower levels, this would be far from sufficient in keeping temperature increases below dangerous levels. It would in fact only delay by a short time the point at which danger thresholds for greenhouse gas concentrations are breached.

It is quite clear from the climate-change science that a new and far stronger agreement is needed. This must be based on the latest science and be negotiated very quickly. This is not straightforward, however.

the detractors

Ever since the first UN climate change treaty was agreed, and despite ever more dire warnings from climate-change science, there has been a strong backlash against taking action to cut greenhouse gas emissions. This has, in part, been led by various industrial groups who have feared that action to limit global warming would impact negatively on their commercial interests. Some coal, oil, and car companies, for example, embarked on campaigns to persuade governments against enacting laws either globally or nationally. This has been especially true in the United States, where massive industry pressure has been applied against government controls on emissions. The United States government has consequently decided to withdraw from the Kyoto Protocol. In Australia, too, coal firms have, among others, prevailed in maintaining official skepticism about the need to agree and implement a new treaty, or even going forward with the existing weak agreements already forged.

The failure to take action on the scale warranted by the latest science has also been partly driven by strong denials from some quarters that there is even a problem. Some commentators have claimed that the world is not warming at all. Others maintain that, even if it is, then it is not caused by human activities but is down to natural factors, such as increased energy output from the Sun. These skeptical voices have been successful in creating an unwarranted level of doubt about the threat posed by climate change, which has been reflected in thousands of media debates that have presented the science of global warming as a two-sided discussion, with arguments both for and against.

The impression this public debate has created is quite misleading, however. On the one hand are the vast majority of climate experts, for example, those working together through the Intergovernmental Panel on Climate Change. This international body, through the collaboration of some 2,500 specialists, is responsible for much of what we now know about climate change and many of the conclusions that they have reached are included in earlier parts of this chapter. Set against this work are the views of a handful of those who deny climate change, some of whom are not even specialists, and some of whom have been funded by oil companies and other commercial interests.

Despite the overwhelming credibility of the mainstream science—and the uncommon level of agreement in the scientific community—many news reports continue to carry a message that suggests serious doubt remains. One random survey of the published scientific reports on climate change found that, of 928 articles published in technical science journals, none doubted that global warming was taking place, or that it was caused by humans. By contrast, a survey of 636 media reports revealed that 53 percent presented some element of doubt. It is certainly vital to maintain and intensify the public discussion about climate change and what to do about it. But it does need to be a properly informed debate.

politics

In addition to resistance to emissions controls that has come from different commercial interests and the doubts that have been aired about the scale of the challenge we face, there have been very complicated international politics to negotiate. This complexity arises from the fact that not all countries are equal in the pollution they create. Some are very poor with tiny emissions per person; others are vast industrial powerhouses. The USA alone is responsible for about one-quarter of greenhouse gas emissions—the single largest creator of pollution—but it refuses to take action, at least any action orchestrated by the national government.

Poor countries say that any obligations on them to cut back pollution would lead to them being less able to reduce poverty. They add that, since they did not create the problem in the first place, it is the rich countries that need to cut back. And they have a point. About two-thirds of the greenhouse gases already in the atmosphere came from a minority of rich countries. Though home to only 13 percent of the global population, these countries have contributed by far the majority of emissions. Even now, despite the rise of China and India, this small segment of global society is still responsible for about half. Some rich countries react by claiming that it would be unfair for them to have to take the lead in cutting emissions: the USA, despite having huge per capita emissions, is one—it claims that it would harm the U.S. economy to cut pollution while China and India did not.

Similar arguments are played out when it comes to deforestation. Countries such as Brazil and India have indicated that they should not be required to protect their own forests for the sake of the global climate as a result of the rich countries cutting down much of theirs centuries ago as part of their own development process. They wonder why they should hold back their development to help solve a problem created by the already developed countries.

Certainly a global deal is needed, but it needs to be a fair one and it does need to reflect the moral responsibility of the rich and better off to go first. Not only are the rich, industrialized countries largely responsible for the warming that is now taking place, they are in a far better position to take action because of the strong economies and comfortable lifestyles enjoyed by most of their citizens.

Fortunately, the commercial interests that have successfully blocked political action on climate change are now being countered by changes in public opinion, at least in some countries. This is leading governments to adopt a different policy. Both the state of California and the UK are now in the process of enacting new legal controls on greenhouse gas emissions that could transform how these societies impact on the global atmosphere. They will not be alone for long. The message is now getting through, and to places where there would formerly have been real resistance.

Encouragingly, some companies are now also pressing for a different approach. In Europe, North America, and elsewhere major firms are calling for official policies and targets to cut emissions. It seems that it is not only the climate that is changing; it is the political mood. And it might be just in time.

what needs to be done now?

If the world is to succeed in avoiding dangerous climate change as the United Nations treaty has set out, cutting back greenhouse gas emissions must be an urgent global priority. The plan that we implement must lead to cuts determined by the latest climate-change science to be sufficient. As already discussed, the consensus is that we should take action to avoid going above two degrees centigrade of average global temperature increase. Although this will still cause major changes, and lead to some loss of biodiversity, it is likely to be at least a manageable level of change for humankind. Going any higher will lead to greater and greater risks, and potentially to a global disaster.

A global plan that dramatically reduces emissions from fossil fuels and that renders land use more compatible with atmospheric stability must be put in place. This must embrace a global program to halt the loss of natural forests and to recreate a large proportion of those already gone.

Until recently it was believed that stabilizing greenhouse gases at about 550 ppm carbon dioxide equivalent might be sufficient to stabilize temperatures at a safe level. We know now that the climate system is more sensitive to greenhouse gas pollution than at first thought, and that we need to stabilize these levels at a point far below this—at a maximum of 450 ppm—and then to reduce these as fast as we can, preferably to below 400 ppm. For industrialized countries, this means setting in place programs that will reduce carbon dioxide emissions by about 80 percent by 2050.

Moreover, we need to start the process of carbon reduction now. If we fail, the action we need to take will be greater and will be less and less likely to happen because it will be even more difficult and costly to undertake. The longer we leave it, the more risky will become the chances of humanity breaching critical atmospheric thresholds.

cut the carbon

Making large-scale cuts in carbon dioxide emissions will require action across a wide range of sectors. One is in power generation. Moving from coal to less-polluting fuels (including some use of natural gas at least as an interim measure) to generate electricity will be necessary. So will the more efficient use of fossil fuels, for example, in so-called combined heat and power systems. These not only generate electricity but also harness heat, for example, to warm homes and offices. This can make the use of gas and other energy sources far more efficient.

The momentum behind industrial economies across the world makes it inevitable that fossil fuel use will continue for some decades to come. To minimize the damage this will cause, it is essential for countries to invest in carbon capture and storage technology. This is a process that removes carbon from fossil fuels and then places it deep in geological deposits, including old oil and gas fields. This technology is new but has huge potential in helping to minimize the climate damage being caused by large-scale fossil-fuel combustion. China has plans to build hundreds more coal-fired power stations that, over their half-century lifetimes, will create vast carbon dioxide emissions. Carbon capture and storage technology could reduce the effects of this considerably.

use renewable power

At the same time as we clean up fossil-powered electricity generation, it is essential that there is a large-scale and rapid expansion of different renewable power technologies. These range from wind to solar power, and from biomass energy (from plants) to wave- and tidal-power technologies. These technologies make only a small carbon contribution, but have the potential to provide clean power indefinitely into the future.

Concentrating solar power uses mirrors to focus the Sun's energy onto pipes containing water. By boiling the water and then using the steam to drive electricity-generating turbines, huge amounts of power can be produced. One estimate suggests that, using just a few percent of the area of the Sahara Desert, this technology could produce power output equivalent to today's total world electricity demand. Large-scale offshore wind farms and tidal schemes could also make a vast contribution.

It is not only on the grand scale that renewable energy technologies can make a difference, great strides can be made in reducing emissions by investing in more small-scale and localized power sources. A range of new technologies are now available that produce power near to where it is used, thus reducing the great losses in electricity that come with long-distance transmission.

our use of energy

How we use energy must alter as well. At present, a lot of the power we generate is wasted in running inefficient appliances, such as the old-fashioned lightbulbs. Switching over to energy-saving alternatives in a wide range of products could make a very substantial difference. Even changing the standby settings in consumer goods that now have little red lights signaling their readiness for use would save substantial amounts of energy.

Normal bulbs (right) use 80% of the energy to produce heat rather than light. Installing energy-efficient bulbs is the first step in Saving Planet Earth.

Changing the design of buildings and the standards that govern their construction would also make a massive positive difference. Many of the buildings that are being erected now will be in use for decades or even hundreds of years. Building to the highest environmental standards now would, thus, provide long-term benefits. In many countries it will also be necessary to embark on major programs of environmental modernization for the existing housing and office stock. Much of this was built before climate-change questions informed design and their inefficient use of power is a major source of carbon dioxide emissions.

Building design has made huge progress in recent years in demonstrating how it is possible to provide housing and work spaces that are not only esthetically pleasing and comfortable to live and work in, but which also cause only a fraction of the carbon dioxide emissions of more traditional buildings. There are also social and quality-of-life benefits that come with super-efficient buildings, for instance, in the form of lower energy bills.

transportation

Our transportation methods will need to become much more efficient as well. New standards requiring that only the cleanest and most efficient vehicles can be sold could transform the contribution from this sector. New, more efficient hybrid vehicles are already in the market and could help slash emissions if they were more widely used. New technologies that could potentially grow in the near future, such as hydrogen fuel cells, also show great promise. Large-scale desert solar-power stations could in the future produce hydrogen, which, in time, could be a viable alternative to gasoline and diesel fuel.

It will also be necessary to look at how we produce and distribute goods, with large pollution savings possible, for example, in producing and consuming food more locally. At present, many people's food is traveling farther and farther to their

Energy Monitor   OUTSIDE TEMP   91 °F

ENGINE

ELEC. MOTOR

Energy | Consumption

tables. Often this is unnecessary and steps can be taken that will sustain good diets while reducing the pollution that comes with it.

Aviation is another sector where we need to reduce emissions. Aircraft are the fastest-growing source of carbon dioxide worldwide and, although not yet at the same total level as power production or cars, are making an increasing impact. And not only are emissions from planes going up, they are also particularly damaging: carbon dioxide and other emissions released at high altitude cause between two-and-a-half and four times as much warming as the equivalent pollution released at the Earth's surface.

There is a certain amount that can be done with technology to reduce emissions from planes, such as using more efficient engines, but improvements are nowhere near matching the

Hybrids are currently the only viable low carbon emission cars.

present rate of growth in the number of flights. If we are to meet the emissions-reductions targets set out by the latest science, we must either cut back even more in other parts of the economy or put policies in place that will at least cut the fast growth in flying that is now taking place right across the world.

International shipping is another area where steps can be taken to cut pollution. This could not only be achieved through changing patterns of production and trade but through modern technologies, including the use of modern sail designs that can be used in tandem with fuel oil in order to reduce emissions considerably. Unfortunately, neither shipping nor aviation emissions are covered under the modest provisions of the Kyoto Protocol, and both therefore remain outside any official international agreement. Putting this right should be a priority for the next round of climate change agreements that will be negotiated in the United Nations.

There is also a need to stabilize and stop deforestation. Alongside some international agreement that will mobilize resources for large-scale conservation programs, there will need to be a stiffening of political resolve in countries that have large areas of forest to halt further forest loss. This in turn will raise complex issues about land ownership and economic development, but those questions must be addressed, and with great determination. Considering that forest loss creates more carbon dioxide emissions than all the world's cars, trucks, and buses, this must remain a top priority for action. Of course, there are simultaneous benefits for biodiversity conservation.

The Alcyone research vessel uses turbosail technology, which cuts fuel consumption by one third.

the moral case

The means to tackle climate change already exist and, if we initiate action right away, it is entirely possible for us to transform the impact we are having on the atmosphere. Certainly technological improvements and breakthroughs that will come in the future will help us do more. We should not wait for any other inventions, however, but should make a good start on emissions reductions right away. Given what we now know about the Earth's climate and how it is changing because of our actions, there is no excuse for delay.

While there is a great deal to be said about climate-change science and the technologies and policies to deal with it, the central point must not be lost: the moral dimension.

The implications of climate change are truly profound: major consequences in biodiversity; a wide range of effects on economic development; and a series of social challenges. Decisions made by societies now will have long-lasting ramifications, determining the nature of life on Earth for thousands of years into the future. We are creating ecological ripples that will reverberate for many generations and we have a moral responsibility to ensure that those ripples are as small and harmless as possible.

Our moral responsibility is not only about the needs of future generations, however. It must also extend to protecting the interests of people who are already born. The poorest and most vulnerable will feel the effects of climate change first and most harshly. Hundreds of millions of people, many of whom have had little to do with causing the problem, will be affected. They have enjoyed very few of the comforts generated by high-carbon lives and yet will suffer from food and water shortages, disease, damage to their homes, and loss of their local natural resources, as, for example, forests and coral reefs succumb to the warming.

Not only do the better off have a responsibility to go first in cutting carbon dioxide and other greenhouse gas emissions in a serious attempt to avoid very damaging levels of global warming, therefore, but we also have a responsibility to help people adapt to the climate change that is already inevitable—see the next chapter for more on this.

Taking the right choices while there is still time is perhaps the greatest challenge that has ever faced humankind. Rising to the challenge and ensuring that our impacts on the planet do not disastrously compromise the climate and, therefore, its people and biodiversity, must be one of our guiding missions for the twenty-first century.

Cantiere per la salvaguardia di San Marco

A rise in the sea level will endanger man-made structures, as well as natural areas. Venice has now built a wall to prevent the rising flood from damaging the city.

whale shark, australia

3. plu

dered
planet

population growth

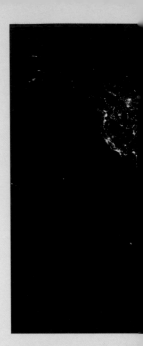

After civilizations were established in the fertile crescent along the valleys of the Tigris and Euphrates rivers, the total human population is estimated to have been around 80 or 90 million people. It grew slowly and, by the time of Christ, is estimated to have reached 200 to 300 million. In the early nineteenth century it is believed that our population reached about one billion. It took until the 1920s for numbers to increase to around two billion and this very significant increase transformed humans into an even more profound ecological influence. More rapid growth followed.

By 1950, less than thirty years later, we had added another five hundred million, pushing our number to about two and half billion. In the forty years after that the human population doubled so that by 1987 it hit five billion people. In October 1999 human numbers are believed to have passed the six-billion mark. Every hour or so, another 10,000 people are born—that's about three per second. It is with good reason, therefore, that our recent increase in numbers is sometimes referred to as a population explosion.

This phenomenal increase in the abundance of the most ecologically important species has been brought about through successive cultural and technological revolutions, among them the emergence of agriculture and the development of technologically advanced urban populations. More effective and more widely available medical treatments have also contributed to a reduction in mortality, thereby increasing our numbers further.

There are counterbalancing impacts from war, disease, and famine, but it is striking that these factors barely register on the upward curve that depicts the growth of human numbers. The last time such an interruption appears to have made a significant impact is during the time of the Black Death, almost 700 years ago. The two world wars of the twentieth century, despite the tremendous loss of human life, created barely a blip on overall growth rates.

There is a vast historical momentum behind the increase in our population, and it is expected to continue for some time yet, particularly in the developing countries. Many of these nations

have a high proportion of children and young people, which means that, when they have their families, and then those children have their own children, our number is set to increase to a total of between eight and twelve billion, and then to stabilize. Estimates change as to the likely foreseeable peak number, but recently researchers have arrived at the figure of about nine billion for the year 2050

As we have seen, the six billion people or so who now live on Earth are causing large-scale changes to the biosphere and atmosphere. This in turn will have serious implications for life, including for humans themselves. Although our ancestors over millennia of prehistory impacted on nature, today our reach is global and has ramifications that will run long into the future.

In meeting our needs for food, energy, water, shelter, and mobility we have already transformed much of the Earth's surface and made profound alterations to many ecosystems. The question our societies must now address is how we can meet the needs of expanding numbers of people while at the same time conserving the planet's biodiversity and maintaining tolerable climatic conditions. This can be done—and it seems that we have little choice other than to achieve this—but it will require us to make changes to how we do things.

The Earth at night. As the world's population continues to expand, more and more lights appear around the world. Only Antarctica cannot be seen.

insatiable appetites for resources

The fortunes of both humankind and the natural world that sustains us are very closely bound together. In order to secure our needs and welfare indefinitely into the future, we need to ensure that the natural world is maintained and that the services it provides for us are protected, nurtured, and sustained.

Demand for the resources that meet our needs continues to increase, however, as the world population rises. Energy demand is expected to be 60 percent larger by 2030. Considering the already far-from-sustainable impacts on the atmosphere arising from the present patterns of fossil energy use (and it is fossil energy that is expected to meet nearly all of the projected increased demand), even keeping this at the present level is not an ecologically viable option, never mind the scale of expanding demand that is anticipated.

A similar, if not larger, increase in demand can be expected for other nonrenewable resources, such as metals. As world consumption of goods from television sets to ovens, and from cars to computers inexorably increases, so the demand for raw materials rises. Other mineral deposits are used to build houses, offices, roads, and airports. Immense quantities of energy and carbon dioxide emissions are linked to the production of cement; there is increased demand for timber, for construction and furniture among other things; and paper use is rocketing upward, placing further pressure on forests, including in the temperate and boreal regions where, from Australia to Scandinavia and from New Zealand to Russia, dwindling stands of old-growth forest are still being felled.

Alongside demand for the gasoline and diesel fuel currently needed to power transportation and farming, we also use huge quantities of crude oil for the manufacture of plastics. Plastics are increasingly used worldwide for products ranging from cellphones to shopping bags, and from water pipes to car bumpers. We also rely on natural gas for purposes other than cooking and power

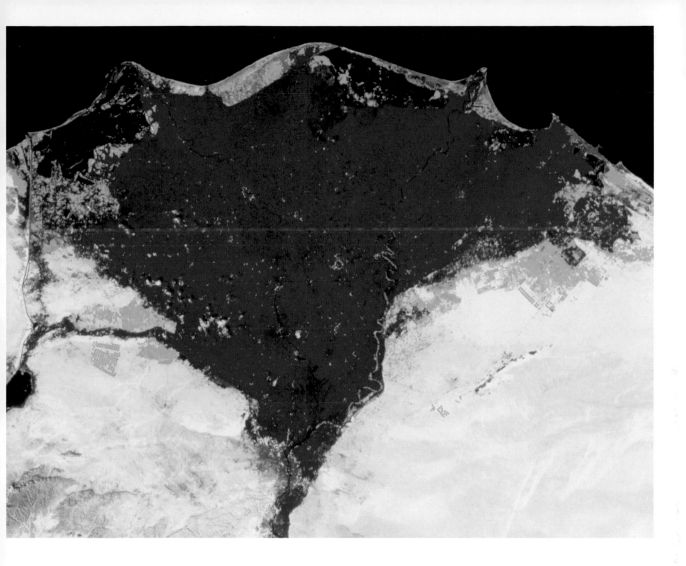

The river delta at the mouth of the Nile River. This false-color image shows how the original fertile land around the delta (red) has now run out, and reclaimed land is now appearing (blue).

generation: one major use is to make the fertilizers that underpin much of modern farming.

Our demand for products and fuel raises many ecological dilemmas. As important as these challenges are, the escalating pressure on another finite resource is perhaps even more pressing. That resource is land—the most precious asset we have.

fat of the land

In many parts of the world the frontier of new agriculture each year encroaches farther into previously natural areas. This expansion includes many of the once densely forested islands of Indonesia.

oil palm plantations

The rapid expansion of oil palm plantations is causing particularly widespread forest loss and, with additional demand for timber and paper, forest loss across Indonesia is among the most rapid in the world. Indonesia has little more than three percent of the world's forests, yet 14 percent of deforestation has recently occurred there. Already half of Borneo's forests are gone, while the amount in Sumatra is more like 70 percent. A further five million acres or so is cleared each year, an area equivalent in size to Wales.

The dense rain forests of the vast islands of Sumatra and Borneo are among the most diverse and unique on Earth: tigers, elephants, rhinos, and orangutans can be found together, the only places on the planet where this is the case. Sun bears, barking deer, clouded leopards, and proboscis monkeys are also among an incredibly diverse fauna, much of which is unique to these islands. Many species are also endangered, including most varieties of the impressive hornbill. Borneo has eight different kinds of these amazing birds; five are regarded as at risk of extinction. The largest ones—the rhinoceros hornbill and the helmeted hornbill—are both declining in numbers, not only because of habitat clearance, but also because they are now unable to breed because the widespread cutting of their large nesting trees.

New species are frequently reported from these islands' forests, which, in a biological sense, are still poorly explored. Yet, in the name of "cheap" food and profitable resource extraction, the extensive rain forest lands are being converted to crop plantations and plundered for their timber.

An oil palm plantation

Each year hundreds of thousands of acres of forests are opened for logging and then for clearance for plantations. The usual pattern is for logging to take out the valuable hardwood timber trees. Where the forest is growing on the islands' extensive peat lands, drains are cut to dry out the ground. The residual forests are either then cut or burned, or both, so as to clear the land ready for plantations. The vast resulting conflagrations have, in recent years, caused a haze of smoke that has spread across thousands of square miles, including over urban areas where it has caused serious public health problems. Often the peat itself catches fire, and can smolder on indefinitely. The oil palm plantations that ultimately replace the once-diverse forest are biologically sterile, leading to the loss of between 80 and 100 percent of the mammals, reptiles, and birds that once lived in the native rain forests.

The area under oil palm cultivation has roughly doubled in the last ten years. The fruits of the oil palm are a prolific and cheap source of vegetable fat that is exported worldwide and used in a huge variety of end products. These include margarine, soup, ice cream, potato chips, bread, lipstick, soap, and cookies. National parks, though set aside for conservation, are not safe as the march of the oil palm plantations reaches new areas.

Other products that are used across the world are also produced in increasing quantities in Indonesia, which places further pressure on the remaining forests. Coffee, for example, is grown across an ever larger area of Sumatra and now impinges on lands that are meant to be protected for biodiversity conservation. A combination of planned and illegal land conversion is thus leading to the very rapid loss of biodiversity in some of the world's most unique and irreplaceable ecosystems.

While the clearance of the forests is sometimes presented as an inevitable aspect of development, it is often the case that the most vulnerable and poorest people pay the highest price. There have been repeated clashes in this area with indigenous and other forest-dwelling peoples whose lands and way of life are being

Forest clearance—the first stage
in the planting of oil palms.

wrecked by the deforestation, and human-rights abuses and disputes over land have been widespread. Even when there are official codes and policies to protect wildlife and local populations these are routinely violated on the back of widespread corruption.

The opening of the forests is leading to more intensive poaching of animals and a rampant illegal trade in endangered species: rare Sumatran rhinos are killed for their horns; tigers are hunted for their skins and teeth; baby orangutans are taken for pets. In fatc, the combination of habitat loss and poaching has led to a precipitous decline in orangutan numbers. One huge fire that raged in central Borneo during 1997 and 1998 is believed—on its own—to have caused the loss of about one-third of the entire island's then-remaining orangutan population. In 2005 the government of Indonesia announced plans for a new and vast expansion of oil palm plantations across central Borneo. Embracing some of the last substantial areas of forest, including protected areas, it will lead to the loss of one of the largest blocks of rain forest remaining in southeastern Asia. Under this pressure, the orangutan could be extinct in the wild within twelve years.

There has recently been added momentum to the spread of oil palm plantations because of increasing global demand for biofuel. Biofuels are plant-derived liquid fuels that can be used as alternatives to gasoline and diesel fuel. Though sometimes presented as a more environmental source of fuel for cars and trucks, if biofuels are produced at the expense of rain forests and lead to the burning of peat, the net environmental effect will be far worse even than using fossil fuel. The rain forests of Indonesia are not the only ecosystems to face this new threat.

Forest fires from space show the extent of forest clearance.

bean feast

A similar situation to that in Indonesia is emerging in Amazonia where soy is not only being grown for food, but increasingly also for fuel. Soy, or soybeans as this food crop is also known, has been cultivated for thousands of years. It was initially grown widely in China and it was not until the second half of the twentieth century that large-scale production got underway elsewhere, starting with the United States.

Soy is an especially important crop because it is so high in protein, acre for acre producing 18 times more than would be produced from the meat of beef cattle. In the Western countries, however, soy is mainly produced to feed animals rather than humans, with the oily part of the grain also used to make margarine, mayonnaise, and sauces.

More recently several tropical and subtropical countries have embarked on large-scale soy production. These include Brazil, where it is one of the recent causes for extensive rain-forest loss.

According to the official estimates, the Brazilian Amazon region lost 10,100 square miles of forest between August 2003 and August 2004, a six percent increase on the previous year. The State of Mato Grosso—at the heart of South America in the southwestern part of Amazonia—showed the greatest loss, with 48 percent of the total deforested area. This is bad enough, but what this figure does not include is the great swathes of savannah (called the Cerrado in Brazil) that is also being lost alongside the tracts of rain forest. The deforestation rate has recently increased even more in Mato Grosso, in large part due to the expansion of soya cultivation.

For decades, soy cultivation in Brazil was restricted to the cooler subtropical regions to the south of the country. As new varieties have become available, however, the area suitable for soy cultivation has expanded: between 1995 and 2004 the land under soy increased by 77 percent. Mato Grosso became the single biggest producer and now soy is rapidly advancing from all sides

Mato Grosso, Brazil in 2002 (top) and 2006 (bottom) showing the rapid deforestation in the area. Most of the cleared land is used to grow soy.

toward the heartland of the Amazon, fueling massive deforestation. While other major factors continue to impinge on the Amazon forests, including cattle ranching and illegal logging, soy production is the primary reason for increased forest loss in recent years.

Soy production is a highly profitable industry, not least because of the vast global demand for animal feed that is, in turn, driven by the increasing number of humans who have meat-rich diets. Amazonian soy can now be an important part in the feed of chickens, pigs, and farmed fish almost anywhere in the world. A massive market exists in Europe and North America, while the expanding middle classes in the supercharged industrial economies of Asia are eating more and more meat too. This itself leads to more pressure to open up forests for lucrative cultivation, and, thus, is leading to more biodiversity loss and the emissions of more climate-changing gases.

As with palm oil, soy is being used to manufacture biofuels. This is in part being driven by concerns about climate change and the need to reduce our dependence on fossil fuels, but is more linked to questions of energy security. Brazil began the large-scale production of biofuels in the 1970s after the world price of oil sharply increased. For decades biofuels were made principally from sugar cane, but soy is now an additional source. Soy oil can also be used to make chemicals, including for example in bioplastics, that are often presented as "environmentally friendly."

Additional demand for soy could accelerate forest loss. While some of the products that will be produced as a result of this might be sold as environmental solutions, the impact on the Earth's atmosphere and biodiversity will be considerable.

There are also implications for human cultural diversity. One tribe, the Enawene Nawe, have been reduced to about 400 surviving members. These people are threatened with cultural annihilation in part through the loss of their ancestral forest homes to fields of soy. Ironically, perhaps, they are one of the few Amazonian tribes who do not hunt for red meat.

The "88" butterfly, photographed in the Mato Grosso rain forest.

the trees for the wood

In addition to being cleared to free land for agriculture, the tropical rain forests are also being plundered to supply the world with wood. Now that the once-extensive rain forests of West Africa, southeastern Asia, and Central America are largely depleted of the most valuable timber species, the loggers are targeting the large blocks of natural forest that remain: in Amazonia, New Guinea, and Central Africa.

Demand for timber drives extensive illegal logging, across much of the Amazon, for example. Indonesia is the world's number one supplier of plywood, but a staggering 80 percent of wood supplied from that country comes from illegal sources, including from supposedly protected areas. In common with the orangutan in Borneo and Sumatra, the loss of Central Africa's gorillas—another of our evolutionary close cousins, and the largest living primates—is being accelerated by the effects of forest exploitation.

The dense jungles of the Central African countries, including Cameroon, the Democratic Republic of Congo, and Gabon are now at the frontline, and are being targeted by logging firms from across the world. The Central African area contains the largest single block of rain forest outside the Amazon basin, but it is rapidly being opened up to gain the valuable resources needed to manufacture wood products such as furniture, doors, and window frames. In order to facilitate the extraction of these resources, logging roads are being driven further and further into previously inaccessible areas.

The logging of timber—and the effects that result from this activity—is now the main pressure on the remaining gorilla populations. Across the Central African rain forests, logging companies have been granted access to 60 percent of the total area. The firms mostly mine the forest for its valuable trees, cutting access roads to get to the preferred stands of timber.

Direct damage to the ape's habitat is one thing, but when the forest is opened up by roads another threat emerges: hunting. Although the gorilla is protected under both national and international laws, the animals are ruthlessly hunted for their meat. A lot of the killing is carried out by the migrant workers who are hired by the logging firms. If, as often happens, they are laid off from work, they sometimes turn to hunting to feed themselves as well as to take the meat to market. An expanding trade in so-called bushmeat, now driven by demand from urban centers where animals caught from the rain forest, including gorillas, are considered as delicacies, is one of the main reasons for the loss of biodiversity across the Central African forests.

The encroachment of people into the once-inaccessible forests is also causing ape diseases to spread, including the deadly Ebola virus. An outbreak of this disease, which can also kill humans, wiped out most of the gorillas and chimpanzees in the Mikembe forest in northern Gabon. The combination of all these factors is leading to an estimated rate of loss of apes in that country of nearly five percent each year. Gorillas are doing better in some protected areas but, in other parts of their range, they have already been eliminated. Across its vast territory in the Central African rain forests, the two species of lowland gorilla (western and eastern) are now variously considered extinct, endangered, or vulnerable, depending on the region in question. Global demand for tropical rain-forest timber, however, is still on the increase while the bushmeat trade thrives.

The destruction of rain forests doesn't only result in the disappearance of animals and plants. Whole ecosystems are destroyed, including humans themselves, at the very top of the food chain. Indigenous peoples, such as this man from South America, exist in delicate balance with their surroundings. Small changes in their culture, such as the introduction of guns or reduction in the areas in which they can hunt, can result in their way of life becoming unsustainable and ultimately lead to them having to leave the area.

It is not only the wildlife and the forest ecosystem that is suffering from the effects of logging. In common with the rain forest peoples of southeastern Asia and Latin America, the human inhabitants of the Central African rain forests are also enduring serious effects. One people who are experiencing terrible consequences are the Bayaka. Their traditional livelihood depends on the forest. That includes hunting, and one effect of exploitation of animals for bushmeat by outsiders is to deprive them of the resources that sustain their survival. The social impacts that accompany the arrival of logging in their lands also include the introduction of diseases, to which they have little resistance, and the accessibility of alcohol, which has led to alcoholism.

As with other tropical regions, the plunder of the Central African rain forests rarely benefits the communities who live in and depend on them. Commercial interests come in search of resources to feed global markets. The huge debt owed by the Central African countries to Western governments and international agencies is a powerful driving force behind the liquidation of countries' natural treasures: in order to earn foreign exchange with which to pay debts, poor countries have little choice other than to export their natural resources. Foreign companies often pay a relatively tiny amount for valuable timbers and minerals, leading not only to massive ecological damage, but also leaving local communities worse off than they were to begin with.

The process of resource liquidation is aided by so-called Structural Adjustment Programs (SAPs) that require nations to open up their economies to foreign firms so that they can benefit from some modest debt relief. In order to continue to pay, however, their natural resources are plundered. Sometimes this is with government backing; at other times the plunder of the trees and land is illegal. Either way, the result is often largely the same: a depleted resource base, lost biodiversity, harm to local communities, and a further contribution of greenhouse gases into the atmosphere.

A combination of hunting and habitat destruction continues to reduce the size of the population of lowland gorillas.

land, people and wildlife

Around the world land is in increasing demand, not only to supply the resources that can fetch lucrative prices in global markets, but also to meet local needs for food and fuel. In many regions the demand for land to meet human needs can conflict with the conservation of biodiversity.

In some countries competing land-use demands can be managed through the establishment of protected areas and the enactment of laws and rules that promote sustainable use of land and wildlife. This may not always be sufficient, however, especially when animals leave the areas set aside for conservation and visit farmland, causing damage to crops. This is widely the case across much of Africa and Asia where, in many rural areas, local communities' crops are visited by elephants in search of food and water. Sometimes this occurs when the animals are on annual migrations; at other times it is because elephant numbers are high and there is insufficient food in natural areas to support them all. Where protected areas such as national parks are small and have a perimeter that is adjacent to farmland, the problem can be severe: it not only causes damage to crops, such as vegetables and melons, but in some cases leads to people being trampled and killed.

A wide range of methods is used to discourage the elephants and to scare them away: from electric fences to loud drum beats, and from bright lights to tear-gas canisters filled with chilli pepper. Some of these methods work, at least for a time, but elephants soon learn they aren't in danger and return anyway.

Namibia, in southwestern Africa, has a unique population of elephants adapted to living in dry conditions. Though they can go for days without drinking by surviving on the moisture in their food, they still need water. This can lead to serious conflicts with farmers and local residents. Thirsty elephants can destroy water holes bored for cattle, and, in the process of getting a drink, the elephants can also wreck fences erected to keep in livestock. Female elephants with small calves are especially destructive in attempting to secure a drink for their young ones, sometimes even trying to get into houses to obtain water.

These kinds of conflicts are not easy to solve. The careful design of protected areas to minimize the contact between crops and large animals is sometimes an option. At times, however, it is necessary to control populations of animals through periodic culls. This is obviously a controversial step but, when all else has failed, it may be necessary to kill some animals in order to retain both a viable population of animals and the wider goodwill of communities whose livelihoods are otherwise on the line.

The demand for resources is not only having an impact on terrestrial ecosystems. Although the consequences are sometimes less obvious, meeting our needs from what we take from the oceans can lead to even more serious conflicts and effects.

The desert elephants of Namibia share their arid home with native peoples. Both have developed ways of making a precarious living in this extreme environment, but both are now suffering from overexploitation of its scarce resources.

hook, line, and sinker

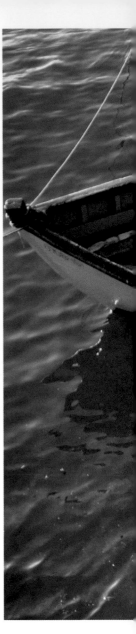

The watery expanses of the great oceans are perhaps the wildest places that remain on Earth. Even out there, however, in the vast and apparently untouched wilderness that covers two-thirds of our planet, the human demand for natural resources is causing large-scale changes. Seafood is hugely popular, nutritious, and healthy, but less positive is the impact that catching fish, lobsters, shrimps, and other species for food can have on the sustainability and biodiversity of the marine environment.

There are now an estimated 3.5 million fishing boats seeking out seafood in the world's oceans. Most are small craft, taking fish from local waters. A small minority, however—about one percent— are classified as large, industrial vessels. This small proportion of boats has a vast capacity for catching fish and is responsible for around 60 percent of the total.

There has been a dramatic increase in the capture of wild marine fish for human consumption. A little over four decades ago the total stood at about 20 million tons. Today it is more than 84 million tons, with more than 40 percent of this catch entering international trade. About 20 million tons a year, approximately 25 percent of all the fish caught, is a "by-catch"—fish that are not commercially valuable and that are put back into the sea, mostly dead. The by-catch does not only include fish, however. It also comprises a wide range of other species, including marine mammals, sea turtles, and sea birds.

As far as whales and dolphins are concerned it is estimated that 60,000 are killed worldwide each year in the fishing gear of vessels. In the North Sea alone it is believed that 7,000 harbor porpoises are drowned annually, mostly in bottom-set gill nets. This is more than two percent of the whole population and is thus a serious threat. Although larger whales can sometimes break free of fishing gear, they can in the process sustain injuries that can lead to illness and starvation.

Turtles, too, suffer from the effects of fishing. Some are caught on baited hooks and others are drowned in trawling gear. Six out

The oceans of the world are some of the most exploited areas of the world. Most fisheries are unsustainable, but even the complete banning of fishing in certain areas has not resulted in the return of the fish. One of the richest fisheries in the world, the cod on the Grand Banks of Newfoundland, was overfished, resulting in the complete collapse and disappearance of the cod in the area. The banning of cod-fishing has still not resulted in the return of the cod there.

Tuna-fishing is now unsustainable in two of the three major populations in the world, all three of which are in international waters beyond the control of individual countries. This has made the imposition of control very hard, but consumer pressure has resulted in the reduction in the by-catch of dolphins.

of the seven species of marine turtle are at present regarded as at some risk of extinction, and the unintentional impacts of fishing are continuing to place these ancient animals at increased risk.

As we eat more fish, many commercially valuable fish species have seriously declined. Cod, tuna, marlin, and swordfish are among those in greatest demand and greatest danger with declines of 90 percent in the most valuable species. As once-productive waters are depleted, fishing boats are going further and further in search of still-healthy populations of their quarry species.

The toothfish became an especially lucrative new target. These fish can reach more than six feet in length, weigh up to 350 pounds, and can sell in some markets for up to $50 a pound. Prices like these have proved an incentive for illegal fishing of this very slow-growing species and rapidly led to the toothfish being added to the list of those that are commercially endangered. The stocks of this fish living off South Africa all but disappeared after just five years of commercial fishing.

Toothfish are among the valuable species sought by long-lining fishing vessels across the great expanses of the southern oceans. Long-line fishing is a technique in which thousands of hooks are attached to lines up to 80 miles long and baited with squid and pieces of fish. These are trailed behind boats to catch as many fish as possible. Aside from the many thousands of tons of valuable fish caught each year with this method, the baited hooks are a lethal attraction for albatrosses and other sea birds, including their close relatives the giant petrels.

Albatrosses and other birds habitually follow boats, and the birds dive onto the lumps of squid and fish as the lines are let out. Albatrosses, being the largest sea birds, inevitably have an advantage over smaller species in the feeding frenzies that accompany prime food being thrown into the sea. The large barbed hooks slice into the bill or throat of the bird and, once the bait is picked up, the bird is slowly pulled under with the weighted line. A vain struggle is soon followed by painful death by drowning. In this way 400,000 albatrosses and giant petrels are killed each year.

It is not only in the southern oceans that sea bird populations are suffering from the effects of long-line fishing. Tens of thousands of sea birds are killed each year in the north Pacific Ocean by long-line fishing boats as well. There are more than 2,500 vessels in the Alaskan long-line fleet; between them they land more than 300 million dollars worth of fish each year. There are also more than 140 long-liners operating out of Hawaii. These fleets between them bait and release 210 million hooks into the sea each year: 210 million chances to catch valuable fish, but also to kill endangered birds.

Information collected by fisheries' officials suggests that the Alaskan long-line fishery kills around 20,000 sea birds each year, including all three species of the North Pacific albatrosses; especially worrying is the loss of critically endangered short-tailed albatrosses. Because of long lining, the black-footed albatross also recently joined the ever-longer list of species threatened with extinction. Of course, the thousands of albatrosses killed by long-line boats do not include the thousands of chicks that starve in the nest after the death of their parents at sea; neither do the figures include the many unknown drowned birds that fall from the hooks before they are dragged back to the boat.

The figures of reported bird deaths are uncertain in part because much of the long-line fishing taking place on the high seas is illegal. In common with the rich financial rewards that can accompany illegal logging, there are strong incentives to break the rules that are intended to conserve fish stocks. A full boatload of toothfish could fetch a boat's crew about $1.5 million, for example. The worldwide value of illegal, unreported, and unregulated catches is vast, and estimated to be between $4.9 and $9.5 billion. Pirate fishing by long-line boats is a major threat to the health of some of the world's fast-dwindling fisheries, but also to some of the most endangered birds.

In restaurants, supermarkets, and stores across the world, the plight of the great albatrosses could not seem more remote, yet it also could not be much closer. Every time we empty a hunk of tuna from a can, or order a portion of halibut, swordfish, or Chilean sea bass, we are directly supporting the plunder that is driving some of the Earth's most charismatic sea birds toward oblivion.

Of the 24 distinct species of albatross recognized today, no fewer than 21 are in danger of being lost for good. While sailors and their mythology have come to treat albatrosses as harbingers of ill fortune, it would perhaps be no surprise if one day we discovered that these amazing birds saw things the other way around.

Black-browed albatross is one
of the 24 species of albatross,
21 of which are in serious
danger of becoming extinct.
They feed by hunting and
scavenging food from the
surface of the ocean,
particularly around fishing
boats. Long-liners
accidentally catch over
400,000 albatrosses a year as
the birds feed on the bait, get
tangled in the hooks and line,
and drown.

the alaskan tundra is one of the most sensitive habitats in the world, where it can be decades—in some cases centuries—before an area returns to normal after disturbance.

nonrenewable resources

The excessive and indifferent plunder of potentially renewable resources, such as fish and timber, is not the only challenge that societies face in achieving more sustainable economies. Our exploitation of nonrenewable resources is causing major impacts as well. For example, mines, pipelines, and oil-production facilities have encroached on many protected areas worldwide, while the pollution produced by these operations has contaminated water and food chains. There have also been many conflicts over land rights with indigenous peoples, who had little if any say about the exploitation of resources found beneath their ancestral lands.

Demand is so strong, however, and the potential financial rewards so great, that these cultural and environmental impacts are often pushed aside in the name of growth, profitability, and development. Though the development gains arising from the exploitation and export of resources are often overstated— certainly in terms of the benefits that will arise for many local people—the case in favor of resource extraction usually prevails, often to the detriment of conservation and sustainability.

It's not just the extraction of resources that is an issue. Processing ores into metals uses vast quantities of energy; the same goes for the manufacture of plastics. More energy is used up in shaping metals and plastics into products. The demand for new products creates demand for new mines, oil wells, and refineries, and locks in more energy use, in turn leading to more pressure on ecosystems and more greenhouse-gas emissions.

how we live

The overall level of demand for resources is ultimately shaped by how we live. And how we live is leading to vast pressure on the natural resource base, not only in respect of renewable products, but also of nonrenewable materials. One study looking at the use of copper estimated that, if everyone in the world consumed goods as people do at present in Western societies, then we would require all of the copper ore left in the Earth's crust, plus all of the copper presently in use, to meet demand for electrical and other goods that need copper. The same study estimated that, of all the copper that was once in the Earth's crust, about one quarter is now lost in nonrecycled waste, for example, in electrical products that are buried in landfills.

Copper mines in several countries have contributed both to the loss of wildlife habitats and to conflicts with local communities. Copper mines have also caused pollution of rivers. Bauxite mines, titanium oxide workings, and iron ore mines have all also been a factor in habitat loss. In winning the means to make aluminum, the white material that is used in some paints and toothpastes and as the raw material for steel, we have respectively generated another set of pressures on land, wildlife, communities, and the atmosphere.

Copper and other metals are not at risk of immediate depletion, but the conclusions that come from the copper study do send an important message about the scale of demand now created by human societies and the extent to which wasteful lifestyles and consumption patterns contribute to impacts on nature.

As we saw in the earlier chapters, it is not only for the sake of the natural world that we should be concerned about the effects caused by our demand for different resources.

The Earth's resources cannot be seen as infinite: the consumption of some metals such as copper is currently unsustainable. We need to both reduce consumption and increase recycling.

for nature's sake and our own

Natural systems, including ecosystems and the atmosphere, provide a range of irreplaceable services, among them the disposal and recycling of waste, purification of air and water, pollination of crops, stabilization of climate, and the provision of nutrients. All of these, and myriad other services provided by nature, are essential for human welfare. Some are of mainly local importance, such as the flow of a stream or fertility of a field; others are of global concern, including climatic stability.

The recent Millennium Ecosystem Assessment looked at the state of ecosystem services as part of a "stock take" of nature. The international team of scientists who undertook the assessment reached some alarming conclusions.

They discovered that, overall, nearly two-thirds (15 out of 24) of the "ecosystem services" that were assessed were found to be undergoing long-term degradation or were being used unsustainably. For example, a high proportion of commercially important fish stocks are overharvested and 15–35 percent of irrigation withdrawals probably exceed replenishment rates. Pressures on ecosystems may be increasing the chance of sudden changes which could harm human well-being, the Assessment said. Examples include new diseases, coastal dead zones, collapsing fisheries, invasive species, and regional climate change.

Perhaps most importantly from the point of view of how countries presently view the process of "development," the team concluded that, unless steps are taken to protect ecosystem services soon, there would be little prospect for ending poverty in the decades ahead. This is not least because it is poorer people who depend most directly on nature and who suffer most as ecosystem services (such as replenishment of soil fertility, provision of freshwater, productivity of fisheries, and supply of fuel wood) decline, degrade or disappear. For many rural poor people, it is nature itself that is their greatest asset.

In better-off societies it is perhaps all too easy to regard our increasingly urban existence as separate from nature and divorced from reliance on the natural world. This is far from true, however. From the fish we buy at the supermarket to the computers in our homes, and from the wooden chairs and tables in our kitchens to the metal shell of our cars, it is on natural resources that we ultimately depend. How we use those resources not only shapes the fortunes of nature, but of people, and especially of the poorest.

This last dimension of the ecological situation we are facing—the differences in consumption levels between richer and poorer societies—is often overlooked. It is, however, essential in shaping how solutions to the challenges of climate change and biodiversity loss must be conceived.

The irrigation of the desert uses water that has lain underground for thousands of years. It is not thought to be replaced, so will eventually run out.

The carbon emissions of this African's
lifestyle is approximately 1/200th of an
American, or 1/95th of a British person.

divided world

In respect of land, energy, and materials, the disparity in resource use between rich and poor is truly vast. While some consumers generate very considerable environmental impacts, for example, through driving large vehicles, taking numerous flights, high levels of product consumption, a meat-rich diet, and high use of gas and electricity, most of the world's population use far less. And yet it is the high-consumption and ecologically destructive lifestyles that are now hard-wired into the culture and economy in many countries, and which are spreading to the wider world.

If the process of development continues to be geared toward extending these Western-style living patterns across the entire world, then it is clear that pressure on the environment and atmosphere will increase. We have already seen how existing demand for resources is leading to serious and alarming change, and it is clear from the present situation that a large-scale increase in resource supply and use would lead to even more rapid and more dangerous regional and global environmental change.

It is increasingly common to hear a view that holds how there is little point in Western countries taking steps to reduce their demand for energy and resources because of the huge and growing impact of the new industrial giants, particularly China and India. It is certainly true that these and other countries have embarked on economic development strategies that have led to a sharp increase in energy and materials. It would be quite wrong, however, for others to conclude that this justifies them carrying on as before.

For a start, compared to what most people in the rich, industrialized countries individually use up, the average per capita use of energy and other resources elsewhere is still reasonably low, even in China and India. There is still widespread poverty in these nations and making an implicit comparison between meeting basic needs there with the profligate waste of resources that takes place in many developed countries is not a fair one. Even if China and India were in any way a legitimate "excuse" for

Developed countries are working on new technology that reduces carbon emissions. This hydrogen-powered London bus only produces water as its exhaust.

Western consumerism, the situation in the truly poor countries surely is deserving of a decisive response by the rich.

According to figures published by the United Nations, the richest 20 percent of the world's population consume 58 percent of the energy, the poorest 20 percent use less than 4 percent of it; the 20 percent of people living in the highest income countries make 86 percent of all consumer purchases, while the poorest 20 percent buy only 1.3 percent of all the products and services sold around the world each year. The disparities are even bigger when it comes to the release of greenhouse gases. In 2003 American per capita carbon dioxide emissions were about 20 tons; citizens from Sierra Leone or Ethiopia (and indeed over much of the rest of Africa) put out barely one-tenth of a ton each. American citizens are each, thus, responsible each year for about 200 times as much greenhouse gas emissions compared to poor African citizens. The UK's per capita carbon pollution contribution, though lower, is about nine-and-a-half tons, still 95 times as much as those in Africa.

The vast inequality in use of resources and emissions cannot be ignored in devising strategies that need to be implemented to slow the loss of biodiversity, limit greenhouse gas emissions, and conserve resources. This is not least because growth in demand for products and energy is inevitable if we are to help end poverty. If that demand is to be met without losing a large proportion of the biodiversity or without causing climate chaos, then it is clear that major reductions in energy and resource use will need to take place in the already rich and industrialized countries. In this sense, the challenge is not so much one of making a choice, or even of finding a balance, between the needs of people and the needs of the environment, it is much more a question of how can we end poverty while at the same time protecting key ecological services.

At present it is often regarded as necessary to sacrifice nature to promote development. Sometimes it is even suggested that moves to protect natural systems are an impediment to ending poverty. The most recent assessment of the state of nature has, however, shown how poverty alleviation programs will falter unless nature is sustained and protected as a prerequisite for development. Any development strategy that does not recognize this basic point will in the end be self-defeating. We rely on the services of ecosystems and climatic stability and, without those, our economies cannot function in the long-term. This is why we need a new and different way of looking at development.

sustainable development

How to lift people out of poverty and to maintain a good quality of life for humanity without ruining the Earth is undoubtedly the great challenge of the twenty-first century. Finding answers to this daunting question is now exercising societies across the world. Fortunately, many are finding that there is an idea that is fit for this historic purpose: sustainable development, or sustainability.

The notion of sustainable development—balancing the social and economic needs of humans with the need to protect the natural environment now and into the future—came to wide attention in 1987 with the publication of the report of the Brundtland commission on environment and development. The Rio de Janeiro Earth Summit in 1992 incorporated the idea into two new international treaties (on biodiversity and climate change) and into a global program for sustainable development called Agenda 21.

Numerous government reports have been published in country after country seeking to put the concept into practice through national action plans. And some progress has been made. But we need to expand the breadth of action if we are to retain this planet's incredible diversity, the benefits the ecosystems provide for us, and the climatic stability necessary for our continued security.

Going further raises some complicated challenges. These include the difficult process of canceling the debts of the developing countries, in setting out new international trade agreements that promote sustainable development, and the means to transfer on a large scale the technologies that will help countries to "leapfrog" past the dirtiest stages of industrial development to those that will not harm the environment. All of these issues have been on the international agenda in recent years, but, so far, too little has been achieved.

Some developed countries have set targets to reduce their climate-changing gases, but most of the reduction is because they are past the dirtiest stages of industrial development. Sustainable development plans aim to allow pre-industrial countries to leapfrog this environmentally damaging stage.

developing countries

These large-scale economic reforms could set the scene for more sustainable development patterns in developing countries where the needs of the local population rather than the demands of the global economy are served first. If livelihoods are secured at the local level, and basic needs met for housing, water, food, and power, for example, then it is expected that population growth will begin to fall and stabilize. Another hugely important factor in reducing population growth rates is education, particularly for girls. Comparatively small sums invested by aid agencies and international development bodies in basic teaching could thus pay planetary dividends. In the end, the most important factors shaping human population growth are economic and aspirational. If people feel more secure, and if they have the ambitions that can come with education, then they tend to have small families. The high fertility rates seen in many developing countries can be partially explained as a necessity—large numbers of people are needed to undertake farm-labor tasks. As needs are met more effectively, and people are less at day-to-day risk, it is not as economically necessary to have more children.

Population growth is a major issued faced by the world. A combination of healthcare and education has resulted in a major change from high to low fertility in Europe, a pattern that can be repeated in other countries.

The transition from high to low fertility that has now happened in Europe will eventually happen across most developing countries, and, as was the case in Europe, it will be linked to better living standards and increased security. The sustainable development challenge is to enable that transition to occur without creating the massive environmental impacts that came with Western industrialization. This in part is about lifestyles and about the use of more efficient and cleaner technologies. It is, of course, not only a challenge for the newly industrializing and developing countries, it is a challenge for all of us. If we are to maintain the planet's biodiversity and climatic stability, then it will be necessary for the high-consuming countries to do even more.

There is no need to wait for any further international agreements; it is possible to make a good start right away. Documents agreed by governments at various international summits have already signed nations up to the principles of sustainable development. And it is not as if the world is bereft of the resources needed to make the transition to sustainable societies. For example, the International Energy Agency says that half a trillion (that is five hundred billion) dollars is needed each year up to 2030 to modernize the world's energy supply. Over more than two decades the world's total expenditure on energy will thus be truly vast. What would be possible if such resources were harnessed for a greener-energy future deliberately geared up to dramatically reduced greenhouse gas emissions?

In many areas of the world, the population depends directly on farming to supply their food. Without environmental justice they will be the first to suffer.

environmental justice

Embedded in the notion of sustainable development is the idea of environmental justice. This works at several levels: it brings moral and ethical dimensions to the sustainability debate and raises questions about people's rights. It is a relatively new idea, but it is increasingly vital to understanding how we might navigate a way forward that will protect biodiversity, conserve resources, and maintain climatic stability.

Serious global environmental-justice issues are raised by the climate-change challenge and the fact that the people who have made least contribution to causing the problem will, on the whole, be its first and hardest-hit victims. Under these circumstances it seems morally correct that the richer countries immediately agree to major cuts in their emissions, and then implement those commitments as a matter of urgency.

Many African farming communities and the societies living on low-lying islands will be among those who will be hit hard by climate change. Natural justice would thus also suggest that rich and industrialized countries mobilize major financial resources to help poorer nations deal with the consequences of climate change that are already inevitable, as well as what is likely to follow from continuing pollution. This needs to come alongside altered development priorities that focus on meeting local needs, including through providing local communities with the means to control their land and other resources more effectively. If successful, such programs might be able to reduce the mass migrations of humans that seem very likely to accompany climatic change.

Other environmental-justice questions are raised through the extraction of resources. For example, it is the tribal peoples of the rain forests who are suffering from the consequences of oil-palm and soybean expansion into their lands, and yet it is the Western consumers and the large food-processing and retailing companies who reap the biggest rewards, in the form of cheap food and elevated profitability, respectively. An approach based on environmental justice would at least seek the consent and active collaboration of communities in land-use changes, and respect

their wishes if they choose to maintain their ancestral lands as forest, and not to convert them to plantations.

Even within the richer, industrialized countries, it tends to be the poorer communities who live adjacent to waste-disposal facilities and to the most toxic sources of industrial pollution. Placing the burden of pollution on the most vulnerable raises moral issues that have yet to be factored into how societies decide to locate the means to sustain high-consuming lifestyles.

solutions or problems?

Even in seeking solutions to environmental challenges, major questions of environmental justice can be raised. For example, in some countries there is now a major planned expansion in the use of biofuels, using corn, soy, and oilseed rape to make gasoline and diesel fuel alternatives. Already the resulting increase in demand for these grains, which had been only used for food and animal feed, is leading to increasing prices for these commodities, and that in turn is affecting the affordability of food for some poor people. To fill the tank of a large car just once with biofuel requires about the same amount of land that is needed to feed one person for a whole year. Is it morally correct to continue with the use of large energy-wasting cars, even if they run on biofuels, or should we reduce emissions and the need for biofuel through the use of much more efficient vehicles?

farming for the future

A key dimension of more sustainable land use must be changes to how people produce and consume food. The globalized model of industrialized farming, which is becoming more and more embedded, is leading to a range of serious environmental and social consequences. We have seen some of these in the previous chapters, from marine dead zones to large-scale greenhouse gas emissions, and from deforestation to conflicts with local communities; much of our food comes with heavy baggage.

More local production for more local markets and producing food with fewer chemicals can help to reduce greenhouse gas emissions and protect biodiversity.

50% of food grown for UK consumption is wasted.

chemicals

Reducing the level of nitrogen used in farming would help to protect biodiversity, conserve fisheries, and help combat global warming. Huge quantities of natural gas are used in the manufacture of fertilizers, while nitrogen oxide emissions that accompany the use of fertilizers are a potent source of global warming too. Given the wide range of very serious impacts arising from excessive nitrogen use considered earlier, artificial-fertilizer reduction will need to be an important aspect of sustainable farming in the future. One way to do this is for governments to implement rules that control the use of nitrogen fertilizers, especially in the most ecologically sensitive areas.

The impact of pesticides can be reduced through using them more sparingly and by the development of more targeted chemicals that break down quickly in the environment. Another approach, which is still the main farming method in many parts of the world, is to use organic techniques. It is sometimes suggested that organic techniques, in producing lower yields per acre on average than some forms of high-input industrial farming, cannot be a major part of future food production, given the expanding demand for food. This is, however, a very limited argument and needs to be put into the context of our wider food economy and, for example, the energy and greenhouse gas savings that can accompany organic methods.

It is also important not to lose sight of how some societies consume food and the impact this has on land use in turn. It is estimated, for example, that 50 percent of the food that is grown for UK consumption is wasted. Some is lost on farms, more in transit, some in the processing, more from the retail stage (with products going past "sell-by" dates, for instance) and then more by the end users, who put a proportion of their groceries into the garbage. This is not only a waste of energy, it also represents an enormous waste of land. And, as we have seen, with a fast-growing human population, land is one of our greatest and most essential resources.

resource management agreements

How we eat and how we farm need to be aligned more with the finite nature of land, biodiversity, and the atmosphere. It can be done, and could create jobs as well as a more secure environment. More sustainable use of the land for farming—using techniques that minimize pollution and that conserve biodiversity—is an essential part of the forward plan. We also need to take steps to ensure that other critical resources are managed in a sustainable manner. Alongside national rules, policies, and controls designed to protect renewable resources and to conserve biodiversity, are various international agreements

sea birds

One that has emerged in response to the impact of fishing on albatross populations is the Agreement on the Conservation of Albatrosses and Petrels (ACAP). This agreement requires the signatories to take steps to reduce sea-bird deaths from long-line fishing, especially by clamping down on pirate fishing boats, which are responsible for about half of albatross fatalities and are, by definition outside any regulation. Australia, South Africa, New Zealand, Ecuador, Spain, and the United Kingdom are the only signatories so far. Given the size of its long-line fleet and the quantity of ocean fish its citizens consume, the United States is notably absent.

Aside from taking action against pirate fishers, a great deal can be achieved by modest adaptations to the methods used by the boats that fish legally. It is estimated that as many as 80 percent of albatross deaths might be avoided by a combination of: weighting lines so that the bait sinks out of reach more quickly; by putting bird-scaring devices onto boats' equipment; and by setting hooks at night. Another approach might involve restricting fishing in some areas at certain times of the year.

Restrictions and mandatory steps, such as requiring boats to fit inexpensive bird-scaring devices and setting lines at night when birds can't see them, have led to a sharp drop in albatrosses killed off the coast of New Zealand by Japanese long-liners, from 4,000

birds per year to fewer than 20. Therefore, it seems that a great deal can be done, and at modest cost.

Changing the gear used, such as the types of hook, can also help boats to avoid taking a by-catch of turtles. Adaptations to shrimp boats' catching gear can help prevent turtles being inadvertently drowned. National laws and international agreements that would make these changes necessary need not entail excessive cost for fishing businesses, and could help reverse the fortunes of some of the world's most critically endangered animals.

Bird-scaring devices can save the lives of 3,980 albatrosses in just one fishery in one year.

logging

In some countries, such as Brazil, Russia, and Indonesia, codes have been put in place to promote sustainable logging of forest ecosystems. But these are, to varying extents, flouted by logging firms more interested in quick financial returns than in the long-term health of the forests. This is especially the case in the developing countries, for instance, Brazil and Indonesia, where illegal logging is rife.

International initiatives too have been put in place to promote more sustainable forestry practices. These have included a target set by the International Tropical Timber Organization to achieve the sustainable management of tropical forests in part through aid money channeled via the Tropical Forestry Action Plan that set out to do the same thing. Despite the repeated commitment of governments to the sustainable use of the tropical forests through these and other initiatives, they are still plundered and mined, rather than carefully sustained to ensure a future harvest both because of widespread corruption in some countries and insufficient enforcement capacity.

The government of Brazil owns 70 percent of the remaining rain-forest area in the country. Over 30 percent of that is included in protected areas, but these sometimes seem only to exist on maps. Across the vast and sparsely populated regions of the Amazon basin, logging firms and others routinely violate lands set aside for biodiversity conservation and for indigenous peoples. This is perhaps unsurprising given the huge value of the natural resources and the limited enforcement capacity of the Brazilian government. A total of 50 Brazilian officials have the job of looking after an area about the size of France.

The situation has not been helped in recent years because of the withdrawal of funding from a number of the richer countries who previously financed some of the forest-conservation efforts that were being put into effect. While, previously, these governments were interested in assisting countries to protect their rain forests, priorities have lately changed. However, if the sustainable use of forests is to be achieved, and if protected areas are to remain actually protected, then many developing countries will need some serious enforcement capacity. And, right now, a lot of them cannot afford it.

A modest contribution in achieving the sustainable use of natural resources can come from various certification schemes. These trace products, such as fish and timber, from where they were harvested and are carried through to the end consumer with an assurance that they were produced in a sustainable manner (or at least more sustainable than many alternatives in the market). Only a minority of products are subject to such schemes, consumer awareness of them is quite low and, if companies choose not to use them, then the problems that are causing the overexploitation of resources remain unaddressed.

resources: for people, or global markets?

extractive reserves

In 1990, Brazil's President José Sarney signed laws allowing more than five million acres of forest areas to be managed by rubber tappers, nut gatherers, and others whose livelihood depended on the harvest of various rain-forest products. The idea was to take an important step toward sustaining the forest by supporting the people who lived in, and on, it. These new land designations were called extractive reserves. They were an ambitious and important measure, and certainly helped to protect the forests from the unsustainable depredations of miners, ranchers, and loggers.

The dynamic of resource exploitation that can be established through mechanisms such as extractive reserves is completely different to that created in large-scale industries that feed global markets—timber, soy, palm oil, and beef cattle among them. These are produced mainly by large corporations, and often sold into global commodity markets. The extractive reserves are under the control of local people who depend in perpetuity on the productivity of the forest, and who thus have a direct interest in sustaining the ecosystems that provide for them and for their communities.

Extractive reserves are of much greater benefit to the local people and the local wildlife, such as this toucan eating fruit, one of the key crops in Amazonian extractive reserves.

land control and legal recognition

It is not only in the forest itself that land control issues are of vital importance. It is also a key question about the lands surrounding the rain forests. Small-scale farmers living on the fringes of forested areas need support and assistance to stay where they are. In many regions, the main cause of continued rain-forest destruction is the migration of people into the forest where they then clear the trees to gain land in order to grow food. This is often presented as an issue linked principally to population growth, but this is misleading—it is the control of land that is the main issue.

In many instances people have been forced to move from land over which they have no title or legal ownership claim, but which has fed them and their families. This is sometimes to make way for export crops grown in large-scale monocultures, such as sugar cane and soy. These crops need lots of land (and water) and, if there is territory occupied by small-scale farmers with no right of ownership, they are often moved on. They then settle in new areas, such as rain forests, leading to further clearance.

Increasing support for small-scale agriculture and stabilizing communities where they already live could help reduce the pressure on the forests. Helping people develop highly diverse agriculture to feed themselves and local markets is one means by which the pressure on the forests can be reduced. This will, however, require governments to review wider economic strategies and the extent to which they see their future in sustainable farming and supplying local markets or in the vast monocultures feeding global demand for cheap bulk commodities.

In addition to helping stabilize farming outside the forests, another vital step is gaining legal recognition of land ownership for the indigenous and other forest-dwelling people already living there. Many of the most promising rain-forest conservation initiatives of recent years have involved land being granted to forest people. In 1990 the Colombian government transferred ownership of half of its Amazonian territory to its indigenous inhabitants in recognition that they would have by far the best chance of conserving the forest in perpetuity. This approach needs to be adopted more widely as any realistic program of large-scale

rain-forest conservation must have local control of resources as a key component.

The loss of habitats like rain forest is very often linked to different kinds of abuses of local people. Control over land is removed from them and either handed out to companies who wish to export resources, or held by central governments who have similar ambitions. In this way, control over the forest and its resources is removed from the people who have the strongest interest in conserving it and handed over to those who have financial interest in removing the most valuable resources, whether that be timber, rare animals, or minerals.

The European invasion of North America in the 1500–1900s resulted in a similar change to the landscape that is happening in South America and South Eastern Asia now, in the form of the loss of indigenous animals and people.

consumption

Changing how resources are managed and who controls them is one side of the sustainable-development equation. Another relates to consumption and how much we are using up. Even if it is possible to develop more ecologically sensitive farming, timber production, fishing, and mining, growing demand would still place pressure on these areas that would lead to unsustainable outcomes.

Though in some high-consuming countries there has been a helpful and constructive discussion about "greener" consumer behavior, this has not, as yet, led to a mainstream debate about overall levels of consumption. However, the volume of what we use is often as important as how it was produced.

In some of these countries there is now a much higher awareness about climate change and the impact of greenhouse gases. This has yet to lead, however, to reduced energy demand. Flying and traffic are both on the increase, while the proliferation of electronic goods continues to keep electricity demand high. At present, nearly all of this demand is met from fossil energy sources. Certainly there is a contribution that must come from renewable energy, but the major gain that can be made immediately is through reduced energy demand.

Given the vast disparity in consumption patterns worldwide, and the fact that successful poverty-reduction strategies will increase demand in what are now the developing countries, it is necessary for reduced demand in the already industrialized societies. A starting point that has been proposed by environmental groups, some companies and governments, is to aim for "factor 4" efficiency gains. This basically means doing twice as much with half the resources, meaning that we would need only one quarter of what we use now. This can be done, and many even argue for a more ambitious factor 10 approach: doing five times as much with half of the resources.

flying is the fastest, but one of most carbon inefficient modes of transportation—one flight in an airplane produces 10 times more carbon per passenger mile than the same distance by train.

recycling

There are a wide range of approaches to take us toward a factor-4 society. One that is familiar to many people is recycling. Recapturing resources out of waste helps reduce pressure on ecosystems because we need to sink fewer new mines and cut down fewer trees. Recycling also saves energy, lots of it. For example, recycling aluminum cans back into new ones only uses about one-tenth of the energy needed to make new cans from scratch out of bauxite ore. Composting of organic wastes can also create a valuable soil conditioner while avoiding methane emissions from landfill.

Recycling and composting is more efficient than so-called energy-to-waste projects, whereby waste is incinerated and electricity generated with the energy released. One recent study found this produced about the same greenhouse gas emissions as a conventional coal-fired power station, not least because of the fossil fuels combusted in the form of discarded plastics.

packaging and materials

Better even than recycling is to avoid waste being created in the first place. In many Western countries packaging is the classic example of unnecessary resource use. Products that are overpackaged create vast quantities of waste, much of which cannot be recycled. As we have seen, the waste of food that is a characteristic of many of these countries also wastes land. Supermarkets that will only sell perfectly shaped fruit and vegetables adds to this, meaning that good-quality produce, though to some eyes misshapen, is rejected.

Product design is another area where great strides toward factor-4 efficiency gains could be quickly made. For example, designing products so they use minimum materials, are durable, use energy as efficiently as the latest technology permits, and are completely recyclable would have a major positive impact on consumption patterns, reducing energy demand and use, and cutting resource waste.

Addressing consumption and, in so doing, reducing demand for resources does not necessarily lead to reduced quality of life, but it does require that societies and people do things differently, and that the consumerist culture that dominates the media and fashion at present is changed.

We can live very well while saving planet Earth, but we cannot continue to have our planetary cake and eat it, as we are now.

Recycling aluminum cans uses 10% of the energy required to produce the same can from the original bauxite ore

planet economics

The heart of the challenge we face in saving biodiversity, conserving resources and avoiding the worst effects of climate change is economic. Most outcomes for the environment are shaped by economic systems and decisions, and at the moment most of these make it economically irrational to protect the planet. For example, in the UK in recent years the real cost of car driving and taking flights has decreased, while the real cost of going by bus or train has increased. The latter two transportation modes are better for the climate, and yet they require an economic sacrifice on the part of the traveler. This is not a good basis for reducing carbon dioxide emissions.

One of the reasons for this failure of economic signals to promote the best choices for our planet is because, at present, we largely don't treat the environment and what it does for us as an economic asset. On balance sheets it is ignored, seen as a free service that will always be there. Whether it is national accounts prepared by governments or a profit-and-loss report from a company, planet Earth is not only treated as if it is economically irrelevant, but damaging critically important environmental systems is actually counted as positive.

The growth of traffic to levels only previously seen in North America and Europe will eventually result in it becoming unsustainable. In North America and Europe levels need to be drastically reduced, while in the rest of the world growth needs to be slowed down and reduced.

gross domestic product

An example of this can be seen in how growth in gross domestic product (GDP) is calculated. GDP is basically a measure of how much economic activity is going on in any particular economy. It is a blunt measure and considers the cutting of ancient forests, plundering of the oceans, and overexploitation of soils, for instance, as contributing to the growth in GDP. Increased greenhouse gas emissions are closely correlated with GDP growth in a number of economies, and even oil spills can count positively toward "growth," because of the money spent on cleaning them up: ships need to be rented, cleaning chemicals bought, and people

employed. This is all good for the economy, but not an accurate reflection of what is happening to the environment.

The failure to reflect environmental values in how we measure the economy is clearly a grave oversight. It is an especially damaging omission considering how economies everywhere depend on nature to provide fundamental and irreplaceable services. Finding ways to correct this anomaly is challenging, however.

putting a price on nature

A conventional economic approach would suggest that the first step is to put a price on nature. Though it is not easy to put a financial value on some aspects of nature, such as the intrinsic value of a species (compared to what it might be sold for) or the beauty of a forest, there have been several attempts to make systematic valuations of nature. One study looked at the question from the point of view of what ecological services would cost to replace. In this review researchers took a range of services provided to us by nature—such as the pollination of crops, purification of water, recycling of nutrients, flood protection, and climatic regulation—and then worked out how much all of that is worth to us.

The 1997 value of the ecosystem services looked at in this study estimated that the annual contribution we gained from nature was about $33 trillion per year. Global GDP in that year was about $18 trillion. This study took a cautious approach to the valuation of ecosystem services, but even it concluded that the services we get from nature each year are far, far larger than the wealth we generate from economic activity. This is a startling finding, and underlines how dangerously out of step our economic thinking is, considering how fundamentally we rely on the normal functioning of the planet for our welfare.

Taking these kinds of findings and then turning them into a different economics that values the planet is clearly an important challenge to which societies must rise. Part of it is about finding a

Without bees, some crops will not be fertilized and will thus not produce seeds, the staple of the world's farming. Bees should be seen to be worth their weight in gold.

means to bring long-term values into short-term thinking. Governments look at the national economy mainly on a short-term basis, for example, in an annual budget. Companies are even more short-term in their thinking, with publicly listed firms often reporting their profits and losses every three months. In both cases, the fortunes of political and company leaders, and, therefore, often their planning, are driven by short-term performance, not the long-term protection of planetary systems.

emissions trading

Steps are, however, beginning to be taken to bring environmental capacities and values into economic decisions. One way is through emissions trading. This is one means whereby a market value can be given to pollutants, or, more accurately, reducing pollutants. Trading schemes have already been successfully employed to cut the sulfur emissions that cause acid rain, and are now being used to help limit carbon dioxide emissions.

The basic idea involves the distribution of a limited number of pollution permits. These are allocated to companies by governments, or better still they are bought through an auction. If a firm pollutes the environment by a lesser amount than the number of permits it has acquired, then it can sell the surplus to those who have not done so well, while companies that pollute more need to buy more permits. Trading schemes can thus create financial incentives to save energy, clean up emissions, and move to renewable power sources.

ecological taxes

Green economic signals can also be sent via ecological taxes. These can be levied on pollution and waste at the same time as tax relief is granted for cleaner and more efficient production and products. Levies on virgin materials can be used to encourage recycling, while taxes on pesticides and nitrogen fertilizers could be helpful in moving toward more sustainable agriculture. Some of the money raised through ecological taxes can be used to meet environmental challenges to, for example, help fund more sustainable farming.

Such tools need to be used alongside a more realistic perception of what the economy is doing for us. To this extent, different indicators and measures are needed so as to paint a more accurate picture of economic performance, including to reflect what is happening to the natural capital of ecosystem services. Economic indicators that measure trends in resource efficiency, biodiversity, and greenhouse gas emissions would enable us to make judgments that are far better informed than those based solely on GDP.

The long-term costs of exploiting the Earth need to be measured to ensure the exploitation is sustainable. Early use of pesticides resulted in huge increases in some crop yields, but early pesticides also resulted in the death of many animals at the top of the food chain.

growth

This does not mean that growth is not a desirable economic goal; in many circumstances it is positive and necessary process. It is, however, the quality of the economic growth that must be measured, not simply by how much it is increasing. If growth is based on the liquidation of natural forests, patterns of damaging resource extraction, high greenhouse gas emissions, serious toxic pollution, and plunder of fisheries, then it cannot be deemed sustainable or economic growth of an acceptable quality. If the main beneficiaries of such growth are the already wealthy, then little benefit will be achieved for poverty alleviation either.

While there has been spectacular growth in the global economy in recent decades, large-scale environmental damage often results. The fact that poverty is still so widespread testifies to how much of the growth that has been achieved has benefited those with already comfortable lifestyles, and done little to improve the welfare of the least well off.

New measures of growth that discriminate between any growth and more positive growth are urgently needed. And it is possible to move quickly in this direction. If done correctly, one possible consequence of different economic policies could be a new industrial revolution, one fit for meeting the challenges of the twenty-first century.

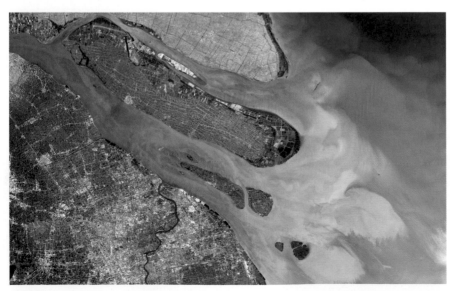

Shanghai, in China, is a city that has recently seen phenomenal growth in population and area covered.

a new industrial revolution

Addressing planetary challenges such as biodiversity loss, resource depletion, and climate change is often seen as posing a threat to development and comfortable lifestyles. While it is increasingly clear that we need to change how we develop economies and meet human needs, the job that we need to do can equally be seen as positive from social and economic perspectives. Indeed, moves toward a more resource-efficient and sustainable economy, which produces far fewer greenhouse gas emissions, could be the basis for a new industrial revolution. New energy-efficient products, waste-treatment technologies, different building designs, new forms of power generation, cleaner vehicles, recycling technologies, new materials technologies, sustainable food production, and the manufacture of water conservation devices are among the new industries that could create wealth and jobs worldwide in the decades ahead.

Such a revolution needs to be stimulated through the creation of incentives, and these need to be put in place by governments and international agencies. These could include new product standards, rules that require large companies to report fully on their environmental performance (thereby highlighting to shareholders and customers where improvements can be made), and through requiring that large financial institutions, such as pension funds, reflect environmental concerns in how they do their business. Alongside measures such as emissions trading and carefully designed environmental taxes, it would be possible to begin a process of industrial transformation quickly.

That process needs to be shared across the world so that countries that are industrializing for the first time can leapfrog the dirtiest and most environmentally damaging stages of development and go straight toward using the most efficient and cleanest technologies available. Aid and international development agencies, international trade and investment agreements, and the official measures used by the industrialized countries to promote exports by their domestic companies could all play a role.

one planet living—the individual

Alongside changes to how societies look at economic development, there are steps that each of us as individuals can make as well. Reducing our individual greenhouse gas emissions is certainly a priority. Most of us are responsible for the annual emission of many, many tons of this most important greenhouse gas, yet we can take immediate steps to cut that "carbon footprint," and thereby our contribution to the climate change that now poses such a grave threat to life on Earth.

Among the biggest energy gobblers are the transformers that continuously recharge your cellphone, power your computer peripherals, and keep your Game Boy ready for use.

energy in the home

A good place to begin is at home. Anything that uses electricity is a potential target. Lightbulbs are an easy start: replacing the old-fashioned incandescent designs with modern energy-efficient alternatives can cut emissions caused by lighting by 80 percent. This is a massive saving with such an easy action, and demonstrates the potential we all have to make a difference. For more expensive products, such as fridges and ovens, seeking out the most energy efficient models available when replacing an old one can lead to further big savings.

How we use energy-consuming products and appliances can be even more important. Switching off lights and other devices powered by electricity when they are not in use saves loads of energy (even if you have already installed the most energy-efficient devices). Unplugging computers and making sure that standby lights are off can make a big contribution to power saving and thus emissions reductions. Eliminating "standby" electricity loss from home appliances could save up to 25 percent on bills, according to a recent study made by the University of California, Berkeley.

Heat is another big source of carbon dioxide emissions. Whether using electricity, gas, or oil-fired boilers, keeping homes and workplaces warm is a huge source of emissions, especially in cooler countries. Saving heat through effective insulation and draft-proofing is thus another effective means of cutting emissions. Turning down the thermostat by just a degree or two can help in some homes, and better-quality doors and windows can pay big emissions dividends too.

transportation

How we decide to get around makes a vast difference to emissions. Using cars less is a step that most people can take. Walking and cycling are often good alternatives for short journeys, while trains can be used instead of cars for some longer trips. More efficient cars that use less fuel and create less pollution are an important way of reducing emissions from the transportation sector. It is, however, generally best to keep an older and less-efficient model of car until it finally reaches the stage where it needs to be scrapped and recycled—there is a substantial body of energy embedded in the old car from its manufacture. Selling it into the used-car market would still mean its emissions continue but the new owner might not keep it running or recycle it. Using the old car less is often the best short-term choice so, when the old car finally reaches the end of the road, seeking out the most efficient model you can (whether new or used) is then a major opportunity to cut emissions. Then, using the new efficient car as little as possible cuts the pollution further still.

In recent years there has been a dramatic increase in flying. Cheap flights, bigger airports, and the convenience of internet ticketing have made flying a far more accessible means of transportation than it was even a couple of decades ago. This increased freedom has come with a huge environmental price, however. There have been some modest technological improvements that have made aircraft cleaner, but this is nowhere near enough to compensate for the growth in flying that continues.

Traveling on planes less frequently is a major step many of us can take. Thinking twice before deciding to take the plane can make a huge difference to an individual's carbon dioxide emissions: many short-haul flights can be avoided by taking a surface transportation alternative, such as a train or a ferry.

food

How we eat also has a huge impact on the state of the planet. Rare oceanic fish, meat reared on a grain diet, which entails intensive farming, and high quantities of dairy produce are among the common food choices that are driving both climate change and the loss of biodiversity. Less of these, and more local food using more sustainable production methods, such as organic, can be important steps toward planetary protection.

Meat is often one of the least effective ways of using limited planetary resources to produce food.

choosing the materials

The materials contained in the vast array of everyday products we use all ultimately come from natural resources. How we consume, therefore, in turn determines levels of mining, plastics production, and forestry operations. By being more discerning consumers, we all play an important part in lessening the pressures on our planet. Looking out for durable products that will last, and then making sure that as much as possible of what we dispose of is recycled, makes an important contribution.

Individual behavior and choices can make a difference, but in our consumer society we need to go much further than voluntary action by individuals in promoting solutions. At present our economy and culture more often encourage people to use more resources than to save them. Advertising rarely embodies an ecological message, and, when it does, it is often overstated in order to give an impression that a product or company is "greener" than it really is. As a result, relatively few people are prepared to change their own behavior to protect biodiversity, conserve resources, or to limit greenhouse gas emissions.

There are limitations to what can be achieved by individuals in the face of rapid planetary change, but it is vital that as many people as possible play their part. This is not just to make a practical and real reduction in our collective impact, it is necessary for individuals to take action so as to change this lifestyle. If more people recycle, ask for energy-efficient lightbulbs, seek out local food, and travel by bicycle, then the culture of societies will start to change as well. This is a vital step, because, if this happens, then the prospects for improvements to policies and laws increase too.

Politicians very often respond to what people want, rather than leading with new ideas themselves. This has been proved time and again, including when it comes to environmental challenges. From the protection of whales, to combating acid rain through to the clean-up of local rivers, and protection of important habitats, most positive environmental improvements have been driven by some form of public demand. More needs to be achieved, however, if we are to effectively address the large-scale challenges that are rapidly unfolding into the twenty-first century.

politics and culture

As more people are becoming aware of the environmental changes taking place on Earth, more and more are taking individual action. This is not only leading to changes to recycling rates, it is helping change the political situation in many countries. Leaders are recognizing that opinions and attitudes are shifting, and that people want to see more action taken to protect the planet. This is making it easier to put in place new rules to make products more energy efficient, to create better recycling facilities, and to gain more support for renewable energy schemes.

In these and many other instances, it is support from government that is the critical factor in bringing about more sustainable societies. Without official backing the transition we need to make probably can't occur, and certainly not in time to avoid major further losses to biodiversity and very damaging levels of climate change. While individuals can do a lot, in the end, it is governments that control the legal processes that can make transformations quickly. For example, it is very important for consumers to buy efficient vehicles, but it is governments who can determine that only the most efficient cars and trucks are being sold.

Governments can further influence the activities of companies, for example, in encouraging them to find out how the resources they use were produced. Was the palm oil in your shampoo produced sustainably by a local community, or was it grown on a massive area of recently cleared rain forest, involving conflicts with local communities, human-rights' abuses, and the burning of peatlands? Most companies that use palm oil in their products don't know the answer to that question, but if we are to have a good chance of saving the rain forests and protecting the climate, it is the kind of information that needs to be collected and acted on by companies. Some firms are trying to find out important information like this, and to change how they work as a result, but the vast majority are not. Governments need to rectify this situation by requiring that all companies collect information about their supply chains and, on the basis of that information, minimize the environmental impacts involved in their businesses.

Governments are also responsible for putting ecological taxes and emissions-trading schemes in place. These potentially powerful tools can only be introduced through official policies and decisions, and, when they are, economic changes begin to occur that have positive impacts across the whole of society. If these and other tools are used effectively by governments it can quickly become cheaper and more convenient for everyone to play their part in reducing pressure on ecosystems and the atmosphere. When that happens, when "greener" behavior becomes the easy and natural thing to do, then dramatic changes quickly follow.

The rapid introduction of lead-free gasoline is a small case in point. Politicians decided it was in the interest of societies to switch to unleaded fuel, and put in place the means to make it happen. Had this been left to individuals buying lead-free fuel as a matter of consumer choice, the rapid and near total phase-out of leaded fuel would probably never have happened—and it certainly wouldn't have happened with the same speed.

Governments also spend hundreds of billions of tax dollars, pounds, euros, yen, and the like to purchase the products and materials needed to sustain public services, from healthcare to transportation systems. If the vast sum of tax money spent at present on securing the cheapest option was instead used only to buy products that meet the best environmental standards, then new markets would instantly be created for more sustainable businesses and technologies. Those technologies would then quickly fall in price as new competitive markets were brought into existence. Protecting the planet is clearly a priority that is in the wider public interest. In fact, it is difficult to imagine more fundamental public good than a stable climate and other vital ecosystem services. This is not, yet, reflected in how governments spend taxpayers' money to procure the means to run public services, however.

Governments are also critical to ensuring that the rules that are agreed, for example, to protect fish stocks and timber supplies, and

Recycling needs to become a legal imperative, not just a moral decision.

to ensure that protected areas are actually enforced. Putting in place the means to monitor the status of natural resources and to ensure that codes are followed is a key role that must be effectively fulfilled in saving planet Earth. Bringing offenders to account can only be done effectively by governments and properly resourced official agencies.

Country leaders and the ministers they appoint are also the only people who can collectively agree new international legal agreements, for example, to protect resources and to reduce greenhouse gas emissions. If there are to be solutions to questions like climate change, then this is an essential collective responsibility of governments. It is complicated, but it is necessary that this collective global responsibility is prioritized by our political leaders.

One major problem, however, is that governments often face controversy and public hostility when they seek to introduce environmental measures. Taxes on fuel and waste, for example, can be politically risky; yet, as we have seen, tools like this are an essential element in the plan for saving planet Earth. The controversy that is provoked by some environmental policies underlines the need for public support for the actions needed to protect nature, conserve resources, and reduce pollution. As individuals we can demonstrate our willingness to change; we can also join in with campaigns run by organizations such as Friends of the Earth that help demonstrate the strength of public demand for the measures needed to protect our planet.

In the end, however, our political leaders will need to show more courage and leadership in making the case for, and then implementing, environmental measures, even if some of the decisions they need to make are considered controversial.

Saving Planet Earth for future generations has now become one of the top political issues, with most European political parties having policies on recycling, sustainable development, and energy efficiency.

a future in paradise

Humankind is faced with a stark choice: continue as now, and destroy much of the planet's biodiversity, climatic stability, and resource productivity, or change course while there is still time and conserve the Earth's incredible ecosystems so they can be used to meet our needs and sustain our welfare indefinitely. While there is increasing demand for measures to conserve the Earth, we are still very far from doing what is needed.

If, starting now and in the coming years, we did begin the transition toward more sustainable societies, our species could put in place the conditions for people to live long and happy lives while protecting our planet and its ability to provide for us indefinitely into the future. We could, if we wanted, maintain the Earth as a productive and vibrant paradise, meeting our needs while conserving its incredible and unique diversity.

Humans are creative and resourceful beings. Over millennia we have solved apparently impossible problems and overcome what seemed like insurmountable challenges. There are several examples from the recent past that show what we can do when we put our minds to it. These and other achievements should give us cause for great optimism.

Smallpox was once a deadly human disease, afflicting our species over generations. For many centuries smallpox seemed a fact of life, a hazard that could wreck communities and mercilessly cull families without warning; its eradication seemed an impossible dream. During the 1970s, however, smallpox was eliminated. Following a planned and properly resourced program of vaccination, the disease was cornered and then finished off. The lives of millions of people were immeasurably improved.

Imagine what could be achieved for sustainable farming and resource use if a comparably well planned and resourced global campaign was put in place.

In the 1940s the world faced the rise of authoritarian regimes that threatened to slash the very fabric of democratic civilization. At first it seemed impossible that sufficient will or resources could be mustered to wage effective campaigns against well armed and determined aggressors. In the USA and the UK, even with only a tiny initial production capacity, it was, however, possible to build up wartime manufacturing very quickly so that vehicles, planes, and munitions were turned out on a vast scale. In many countries, people volunteered to join the collective effort of resisting, and then defeating, a common enemy. Political leadership, public support, and technology came together quickly and succeeded in stopping totalitarian brutality on a scale never before seen.

Imagine what we could do to limit greenhouse gas emissions with that kind of commitment and mobilization of technology and industry.

At the turn of the twentieth century many people doubted that it was possible to build flying machines that were heavier than air. Then, in 1903, the Wright brothers lifted off in their primitive plane. Even with this breakthrough, very few people believed that it would be possible in the same century for people to develop the means to travel to the moon and back. In 1969, however, people did just that. A major government program of research and development overcame one technical barrier after another, until it was possible to meet the challenge at hand.

Imagine what could be done to protect species and habitats with a high-profile, well resourced program.

It is this scale of ambition that is needed now to save planet Earth. Societies, with the backing of individuals, and with leadership and technical innovation from governments and companies, need to begin the transitions required to secure ecosystem services that we all ultimately depend upon. The program we need must range from the establishment of sufficient

In 1961, President Kennedy mobilized the United States of America to send a man to walk on the moon before the end of the 60s: it took nine years to do what most people thought was impossible. The saving of planet Earth requires just such an effort, but is easily within the abilities of the human race.

protected areas through to the introduction of ecological taxes and embrace technologies that produce renewable electricity and eliminate the waste of materials. It will require the forging of complex and, in some ways, controversial international agreements and the planned and enforced sustainable use of a wide range of resources and ecosystem services.

Our planet can support nine billion people, but not with everyone living as many people in the high-consuming industrialized countries do at present. To provide for expanding human needs we will need to change consumption patterns. This needn't be a negative process, however. It is possible to secure comfortable lives while dramatically reducing the pressures we are placing on our unique and irreplaceable home planet.

If people signal that this is what they want, then it can happen. If, in contrast, we carry on as we are then serious changes that will lead ultimately to threats to our own security are inevitable.

We have the means to live in paradise, and to establish the conditions in which future generations can do the same. Now we just need to do it. The power to do so is in many ways with you, the reader. Do you want to save planet Earth?

picture credits

index